Xuezhong Wu · Xiang Xi · Yulie Wu ·
Dingbang Xiao

D1824668

# Cylindrical Vibratory Gyroscope

National Defense Industry Press

Xuezhong Wu
Changsha, China

Yulie Wu
Changsha, China

Xiang Xi
Changsha, China

Dingbang Xiao
Changsha, China

ISSN 2195-9862          ISSN 2195-9870  (electronic)
Springer Tracts in Mechanical Engineering
ISBN 978-981-16-2728-6          ISBN 978-981-16-2726-2  (eBook)
https://doi.org/10.1007/978-981-16-2726-2

Jointly published with National Defense Industry Press
The print edition is not for sale in China (Mainland). Customers from China (Mainland) please order the print book from: National Defense Industry Press.

Translation from the Chinese language edition: *Yuan Zhu Ke Ti Zhen Dong Tuo Luo* by Xiang Xi, et al., © National Defense Industry Press 2018. Published by National Defense Industry Press. All Rights Reserved.

This Springer imprint is published by the registered company Springer Nature Singapore Pte Ltd.
The registered company address is: 152 Beach Road, #21-01/04 Gateway East, Singapore 189721, Singapore

# Springer Tracts in Mechanical Engineering

Springer Tracts in Mechanical Engineering (STME) publishes the latest developments in Mechanical Engineering - quickly, informally and with high quality. The intent is to cover all the main branches of mechanical engineering, both theoretical and applied, including:

- Engineering Design
- Machinery and Machine Elements
- Mechanical Structures and Stress Analysis
- Automotive Engineering
- Engine Technology
- Aerospace Technology and Astronautics
- Nanotechnology and Microengineering
- Control, Robotics, Mechatronics
- MEMS
- Theoretical and Applied Mechanics
- Dynamical Systems, Control
- Fluids Mechanics
- Engineering Thermodynamics, Heat and Mass Transfer
- Manufacturing
- Precision Engineering, Instrumentation, Measurement
- Materials Engineering
- Tribology and Surface Technology

Within the scope of the series are monographs, professional books or graduate textbooks, edited volumes as well as outstanding PhD theses and books purposely devoted to support education in mechanical engineering at graduate and post-graduate levels.

**Indexed by SCOPUS, zbMATH, SCImago.**

Please check our Lecture Notes in Mechanical Engineering at http://www.springer.com/series/11236 if you are interested in conference proceedings.

To submit a proposal or for further inquiries, please contact the Springer Editor **in your region**:

Ms. Ella Zhang (China)
Email: ella.zhang@springernature.com
Priya Vyas (India)
Email: priya.vyas@springer.com
Dr. Leontina Di Cecco (All other countries)
Email: leontina.dicecco@springer.com

All books published in the series are submitted for consideration in Web of Science.

More information about this series at http://www.springer.com/series/11693

# Preface

Gyroscopes are devices which use inertia to sense the rotation of moving objects. They are widely used and play indispensable roles in many fields, including those of aerospace, aviation, navigation, and weaponry. In the 1940s, the aviation industry made extensive use of a mechanical gyroscope supported by plain ball bearings. Since then, gyroscope systems have been constantly updated, with their performance continuously increasing and volume increasingly shrinking. At present, there are a number of different types of gyroscopes in the market, including mechanical gyroscopes, optical gyroscopes, vibratory gyroscopes, and gyroscopes based on other new principles. For the foreseeable future, the sustainable growth of the market for gyroscope applications will continue to promote the continuous, long-term development of gyroscope technologies, and shell vibratory gyroscopes form an important subset of these technologies. They have been proven to have prominent advantages, including high accuracy, small size, considerable stability, and formidable impact resistance, and this makes them highly applicable for use in precision-guided weapons, especially in tactical weaponry. This in turn means that these gyroscopes possess bright market prospects. Therefore, the development of the shell vibratory gyroscope is of great practical significance for improving the independent R&D and innovation capabilities of the inertial navigation industry.

Although it is simple in structure, the design and manufacturing technologies used in cylindrical vibratory gyroscopes (CVGs) come from many fields of knowledge, including the theory of plates and shells, structural dynamics, vibration mechanics, precision machinery manufacturing, and control theory. The authors of this book have made full use of their years of scientific research experience and the development of practical applications in order to systematically explain the technological systems of CVGs from six aspects, including CVG working mechanisms, theoretical modeling, kinetic analysis, manufacturing engineering, test methods, and control strategies. Chapters 2–4 review the foundational theory of CVGs and construct a systemic theoretical model for the CVG. Chapters 5 and 6 explain the manufacturing techniques used to produce the cylindrical resonator casings, which serve as the foundation for producing high-performance CVG devices. Chapters 7 and 8 explain methods of circuit control and compensation in gyroscopes, primarily including typical self-oscillation-driven circuits and force-to-rebalance control circuits, as well as a compensation method designed according to the unique characteristics of CVGs. This book also presents some of the latest achievements in CVG development, thereby allowing the readers to quickly understand and master basic knowledge of the CVG as well as the latest progress in this field.

The book is intended to serve as a professional and technical reference for people engaged in the research of gyroscopes and inertial navigation systems. It can also be used as an elective textbook for senior undergraduates majoring in machinery and navigation, and can provide them with an additional understanding of inertial navigation devices.

The book was authored by Wu Xuezhong et al. The authors would also like to thank Tao Yi, Zhu Bingjie, Xie Di, and Zhang Yongmeng, and all postgraduates at the Micro-Nano System Laboratory of the National University of Defense Technology, for providing the book with important contributing materials and suggestions. We would also like to thank postgraduates Qu Luozhen and Sun Jiangkun for their strenuous efforts on some of the chapters.

The book also provides a summary of the authors' work on a key project supported by the National Natural Science Foundation of China (51335011 and 51935013), and we should also like to extend our heartfelt thanks to the National Natural Science Foundation of China for their long-term funding and support.

Considering the tight deadline for publication and the authors' limited knowledge, the inclusion of some careless mistakes and errors in this book is regretfully unavoidable, and we respectfully invite our readers to share their critiques and opinions with us.

Changsha, China
April 2017

Xuezhong Wu
Xiang Xi
Yulie Wu
Dingbang Xiao

# Contents

# Chapter 1
# Introduction

## 1.1 Foreword

An inertial navigation system (INS) is an autonomic navigation system that does not rely on the input of outside information, and does not radiate energy to the outside world or easily become subjected to outside interference. INS systems have numerous military applications, and are frequently used in precision-guided missiles, military aircraft, surface ships, ground vehicles, and personal military equipment, and play a decisive role in determining the tactical and technical properties of various military systems, including the properties of accuracy, reliability, maneuverability, and speed.

A gyroscope is the key component of an INS and directly determines the system's cost and performance. Depending on the working principles of a given gyroscope model, the most currently used gyroscopes can be divided into three categories: (1) mechanical rotor gyroscopes, such as liquid floated gyroscopes, dynamically tuned gyroscopes, air-supported gyroscopes, and electrostatic gyroscopes; (2) Coriolis vibratory gyroscopes, such as hemispherical resonator gyroscopes (HRG), cylindrical piezoelectric gyroscopes, MEMS (micro-electromechanical systems) resonant ring gyroscopes, and tuning-fork gyroscopes; (3) optical gyroscopes, such as laser gyroscopes, optic fiber gyroscopes, and integrated optic gyroscopes. According to the magnitude of a given gyroscope's bias stability, these systems can also be classified into the following categories: (1) inertial gyroscopes, which have a bias stability of less than 0.01°/h, and are primarily used for high-precision navigation in spacecraft, large ships, and submarines; (2) tactical gyroscopes, which have a bias stability ranging from 0.1 to 10°/h and are primarily used for medium-precision navigation in civil aircraft, satellites, and medium/short-range missiles; (3) rate gyroscopes, which have a bias stability greater than 10°/h and are primarily used for short-term or low-precision navigation in low-cost robots and vehicle attitude sensors.

The cylindrical vibratory gyroscope (CVG) covered in this book is a prominent type of Coriolis vibratory gyroscope. It performs angular velocity detection by using the Coriolis effect as measured through the standing-wave vibrations of the revolving resonator shell. As CVGs have no wearable components, they have a long service life.

© National Defense Industry Press 2021
X. Wu et al., *Cylindrical Vibratory Gyroscope*, Springer Tracts in Mechanical Engineering,
https://doi.org/10.1007/978-981-16-2726-2_1

In addition, due to their high structural symmetry, vibratory shell gyroscopes also possess prominent advantages such as high precision, a large operating temperature range, rapid activation, and a lack of sensitivity to shock overload [1]. As such, they have received widespread attention from the industry in recent years.

## 1.2   Overview of Coriolis Vibratory Gyroscopes

A Coriolis vibratory gyroscope is a type of rotorless gyroscope. A vibratory element is used for angular velocity detection in place of traditional interior mechanical rotors. At a gyroscope technology seminar held in Germany in 1998, American gyroscope expert D. D. Lynch pointed out that Coriolis vibratory gyroscopes not only had all the inertial qualities of the other two types of solid-state gyroscopes (laser gyroscopes and optic fiber gyroscopes), but were also smaller in size [2]. At a later date, the IEEE GAP compiled a specification guide and testing procedure for Coriolis vibratory gyroscopes. At present, Coriolis vibratory gyroscopes have been named a new type of solid-state gyroscope, with great development potential, and are drawing significant attention from the international inertial technology community.

Coriolis vibratory gyroscopes can be divided into a number of different categories depending on their characteristics: by base materials, into the categories of silicon and non-silicon gyroscopes; by drive modes, into categories of electrostatic gyroscopes, electromagnetic gyroscopes, and piezoelectric gyroscopes; by sense modes, into categories of capacitive detection gyroscopes, piezoresistive detection gyroscopes, piezoelectric detection gyroscopes, optical detection gyroscopes, and tunnel effect detection gyroscopes; by operating modes, into categories of rate gyroscopes and integrating gyroscopes; by machining methods, into categories of gyroscopes made by bulk micromachining technology, gyroscopes made by surface micromachining technology, and gyroscopes made by LIGA. Generally, they can also be sorted in terms of their vibration structures, either into the categories of vibratory beam gyroscopes, tuning fork gyroscopes, shell vibratory gyroscopes, or vibratory plate gyroscopes.

### I.   Vibratory beam gyroscopes

A micro-gyroscope with a vibratory beam uses a straight beam as its vibratory element, and its typical structure is shown in Fig. 1.1. The central axis of the beam is the input axis of angular velocity, and angular velocity input is implemented by sensing the beam's vibration. The most common driving and sense modes in gyroscopes of this structure are piezoelectric driving and piezoelectric detection. A prominent representative product of this type is the Gyrostar, which was developed by Murata, a Japanese company. This gyroscope has a simple structure, works reliably, and is easy to manufacture, but since its sense mode is excessively coupled with its drive mode, detection is difficult to perform and its sensitivity is low.

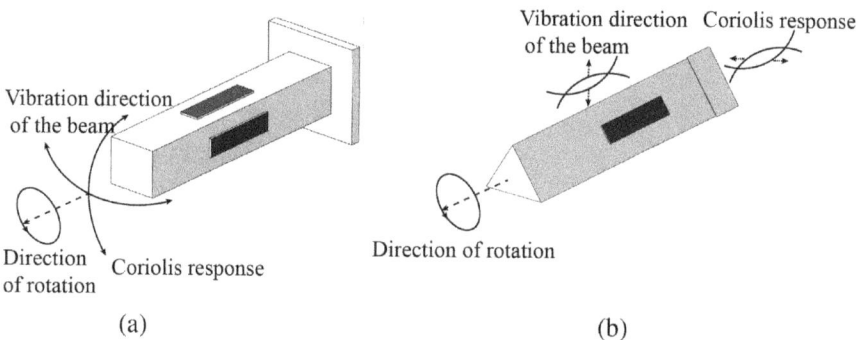

Vibration direction  Coriolis response
of the beam

Vibration direction
of the beam

Direction
of rotation   Coriolis response

Direction of rotation

(a)

(b)

**Fig. 1.1**  Vibratory beam gyroscopes **a** rectangular beam structure; **b** triangular beam structure

## II.  Tuning fork gyroscopes

The working principle of a tuning-fork micro-gyroscope (Fig. 1.2) is similar to that of a micro-gyroscope with a vibratory beam, except that the former is equipped with tuning forks as its vibratory element. In general, they use a single quartz wafer and etch it into an entire tuning fork and supporting structure, while the oscillator drives the tuning forks to move in opposite directions. When angular velocity is applied, the tuning forks will be affected by the torque of the Coriolis force, generating vibrations in sense mode. The structure separates the gyroscope's drive mode from its sense mode, making it easier to detect signals. BEI Systron Donner, an American company, specializes in developing gyroscopes of this structure.

## III.  Shell vibratory gyroscopes

The most prominent feature of vibratory shell gyroscopes is their centrosymmetric revolving shell structure (Fig. 1.3). This may be a hemispherical structure, a cylindrical structure, or a ring structure. In this type of gyroscope, any two mutually orthogonal axes on the cross section of the vibratory shell can be used as a driving axle and sensory axis. A hemispherical resonator gyroscope (HRG) consists of three parts, including a hemispherical resonator, an excitation hood, and a sensory base. An

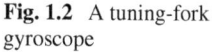

**Fig. 1.2**  A tuning-fork gyroscope

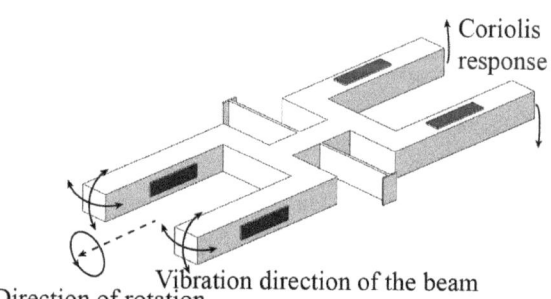

Coriolis response

Vibration direction of the beam
Direction of rotation

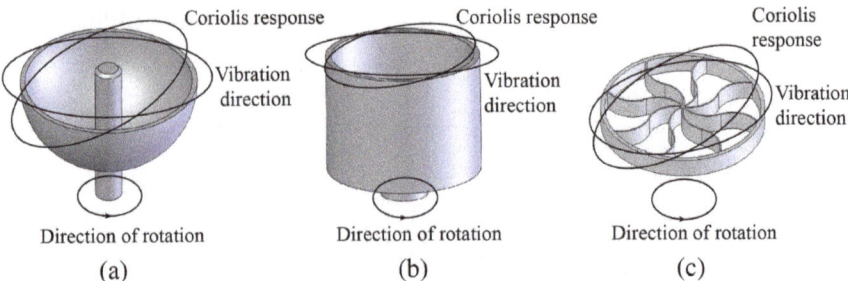

**Fig. 1.3** Shell vibratory gyroscopes **a** Hemispherical structure; **b** cylindrical structure; **c** ring structure

electrostatic force is used to drive the hemispherical resonator to generate standing-wave vibrations. After changing the capacitance of the base surface and sensing the internal surface of the hemispherical resonator, resonator displacement can be calculated to reveal the gyroscope's rotation angle [3]. Silicon Sensing Systems, a Japanese company, has developed an electromagnet-driven ring vibratory gyroscope that can be used for electromagnetic detection. This gyroscope has a permanent magnet on its microstructure, and when currents flow through the electrified gimbal, a force is generated to excite the ring structure, while Coriolis force causes the ring structure to move, driving the gimbal to cut across the magnetic field, thus generating voltage to detect angular velocity.

IV.    Vibratory plate gyroscopes

Although the vibratory element of a gyroscope with a vibratory plate (Fig. 1.4) uses a plate structure, gyroscopes of this structure have developed rapidly due to their great possible variations in plate shape, vibration mode, and operating mode. Leading developers include Draper Laboratory (USA), Analog Device (USA), and Sensonor (Norway). A vibratory wheel gyroscope designed by U.C. Berkeley (Berkeley Sensor and Actuator Center) and manufactured using surfacing technology can be used to detect angular velocity input in two different directions by using a disc structure.

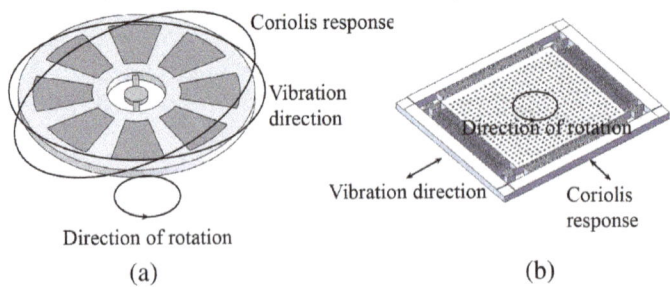

**Fig. 1.4** Vibratory plate gyroscopes. **a** A linear vibratory disc structure; **b** a linear vibratory plate structure

Chapter 2 introduces the working principle and structure of CVGs. In specific terms, this chapter mainly introduces the basic structural types and working principles of CVGs, describes the basic principle of piezoelectric driving and detection, and presents dynamic and electrical models of piezoelectric actuators and sensors.

Chapter 3 presents a theoretical analysis and modeling of CVGs. Chapter 3 provides a theoretical analysis and modeling of CVGs, and presents a basic mathematical model of CVG resonators. By analyzing the radial stiffness and axial stiffness of ring resonators, this chapter describes the resonator's concentration stiffness model and concentration mass model.

Chapter 4 provides a kinetic analysis and modeling of CVGs. This chapter introduces the kinetic equations of CVG resonators, sets forth the applications of the assumed mode method, dynamic magnification method, and mode superposition method in the resonator's dynamic response analysis. It continues by deducing the Coriolis moment of gyroscopes under the action of external angular velocity and calculates their steady-state response in sense mode as well as the output of detection signals in this mode; the chapter also presents a sense model for measuring angular velocity.

Chapter 5 is concerned with the manufacturing of CVGs. This chapter introduces the relationship between the manufacturing errors of resonators and their dynamic performance. The chapter presents a general machining process of metal resonator structure according to the machining characteristics of thin-walled components made of constant modulus alloy, and presents a mass balancing technique for Imperfect Resonators.

Chapter 6 covers parameter testing for CVG resonators. This chapter introduces methods for testing the major parameters of machined gyroscope resonators, including a frequency characteristic test, quality factor test, mode shape test, and piezoelectric electrode parameter test.

Chapter 7 concerns the full-closed loop control of CVGs. Based on electromechanical coupling characteristics of CVGs, this chapter presents an equivalent circuit model of resonators and a method for identifying equivalent circuit parameters. It continues by covering phase control-based resonant excitation methods and fixed amplitude control techniques concerning the stability of driving and vibrations, and introduces force-to-rebalance techniques for resonators along with methods for the demodulation and test of angular velocity.

Chapter 8 introduces the issues of performance error and compensation in CVGs. It mainly introduces error sources and error models of CVGs, and some practicable compensation methods are proposed for temperature errors in gyroscopes.

# References

1. Yang, Y. F., & Zhao, H. (2009). *Solid wave gyros*. Beijing: National Defense Industry Press.
2. Lynch, D. (1998). Coriolis vibratory gyros. In *Proceedings of the symposium gyroscope technology*, Stuttgart, Germany.

3. Izmailov, E., Kolesnik, M., Osipov, A., & Akimov, A. (1999) Hemispherical resonator gyroscope technology. Problems and possible ways of their solutions. In *Proceedings of the RTO SCI international conference on integrated navigation systems.*
4. Rozelle, D. M. (2013). The hemispherical resonator gyro: From wineglass to the planets. *Advances in the Astronautical Sciences, 134,* 22.
5. Watson, W. (2008). Vibratory gyroscope skewed driver and pick-off geometry. *IEEE Sensors Journal, 1,* 332–339.
6. Anders, J. T., & Pearson, R. (1994). Applications of the start vibratory gyroscope. *GEC Review, 9*(3), 168–175.
7. Chikovani, V. V., Okon, I. M., Barabashov, A. S., & Tewksbury, P. (2008). A set of high accuracy low cost metallic resonator CVG. *IEEE/Ion Position, Location and Navigation Symposium, 1–3,* 95–100.
8. Watson, W. S. (2010). Vibratory gyroscope skewed pick-off and driver geometry. In *Proceedings of the position location and navigation symposium (PLANS),* 2010 IEEE/ION. IEEE.
9. Watson, W. S. (2000). Vibrating structure gyroscope performance improvements. In *Proceedings of the symposium gyroscope technology 2000,* Stuttgart, Germany.
10. Cho, J., Gregory, J., & Najafi, K. (2011). Single-crystal-silicon vibratory cylindrical rate integrating gyroscope (CING). In *2011 16th international on proceedings of the solid-state sensors, actuators and microsystems conference (TRANSDUCERS).* IEEE.
11. Cho, J., Yan, J., Gregory, J., Eberhart, H., Peterson, R., & Najafi, K. (2013). High-Q fused silica birdbath and hemispherical 3-D resonators made by blow torch molding. In *Proceedings of the Proc IEEE MEMS.*

# Chapter 2
# Working Principles and Structure of CVGs

The CVG is a type of shell vibratory gyroscope which functions by exciting and detecting standing waves. They are equipped with a cylindrical resonator, and their typical working modes include piezoelectric drive and piezoelcctric sense mode. This chapter begins with a brief theoretical description of the Coriolis effect present in vibratory gyroscopes, then introduces the working principles and overall structure of the CVG.

## 2.1 Coriolis Effect in Vibratory Gyroscopes

A vibratory gyroscope generally has one or more vibrating components, and its angular velocity is detected by using the Coriolis effect [1]. The concept of Coriolis acceleration was first proposed by French scientist G. G. de Coriolis (1792–1843). All motions are relative, so a given motion varies between different frames of reference. The relationship between the motions of objects in different frames of reference is known as a composite motion problem. In kinetics, the selection of a frame of reference is of fundamental significance, because Newton's laws can only be applied in an inertial frame of reference. Two frames of reference are selected in the space where a particle is moving. One is a fixed frame of reference, which is set to an inertial coordinate system, while the other is a moving frame of reference set to moving coordinate system o-xyz, as shown in Fig. 2.1. If the moving coordinate system is in rotational motion and translational motion relative to the inertial coordinate system, the state of motion of the particle relative to the inertial coordinate system depends on the motion of the particle relative to the moving coordinate system as well as the motion of the moving coordinate system relative to that of the inertial coordinate system.

Let the displacement, velocity, and acceleration of the particle in the inertial coordinate system be $\vec{r}_i$, $\vec{v}_i$, and $\vec{a}_i$; let its displacement, velocity, and acceleration in the

© National Defense Industry Press 2021
X. Wu et al., *Cylindrical Vibratory Gyroscope*, Springer Tracts in Mechanical Engineering,
https://doi.org/10.1007/978-981-16-2726-2_2

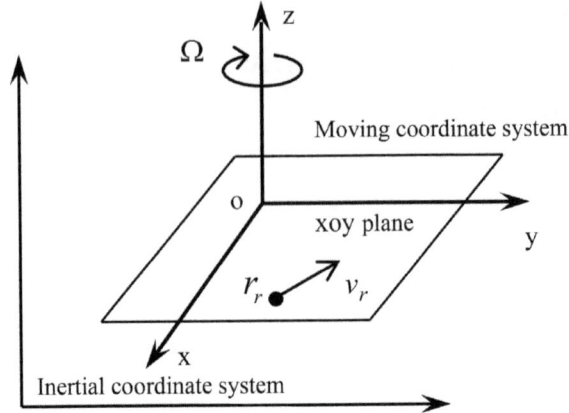

**Fig. 2.1** Rotational dynamic system in the inertial frame of reference

moving coordinate system be $\vec{r}_r$, $\vec{v}_r$, and $\vec{a}_r$; let the displacement, velocity and acceleration of the moving coordinate system relative to the inertial coordinate system be $\vec{r}_o$, $\vec{v}_o$, and $\vec{a}_o$, and the rotational angular velocity be $\vec{\Omega}$. Therefore, the displacement, velocity, and acceleration of the particle in the inertial frame of reference are as follows:

$$
\begin{cases}
\vec{r}_i = \vec{r}_r + \vec{r}_o \\
\vec{v}_i = \frac{d\vec{r}_i}{dt} = \frac{d\vec{r}_r}{dt} + \frac{d\vec{r}_o}{dt} + \vec{\Omega} \times \vec{r}_r \\
\vec{a}_i = \frac{d\vec{v}_i}{dt} = \frac{d^2\vec{r}_i}{dt^2} = \underbrace{\frac{d^2\vec{r}_r}{dt^2}}_{\vec{a}_r} + \underbrace{\frac{d^2\vec{r}_o}{dt^2} + \frac{d\vec{\Omega}}{dt} \times \vec{r}_r + \vec{\Omega} \times (\vec{\Omega} \times \vec{r}_r)}_{\vec{a}_e} + \underbrace{2\vec{\Omega} \times \frac{d\vec{r}_r}{dt}}_{\vec{a}_c}
\end{cases}
$$

$$(2.1)$$

This is a vector composition formula for the state of particle motion. The left side of the third equation is the absolute acceleration of the particle, while the first term on the right side of the equation is relative acceleration $\vec{a}_r$. The second, third, and fourth terms on the right are referred to as the convective acceleration $\vec{a}_e$ while the fifth term is known as Coriolis acceleration $\vec{a}_c$. That is, the acceleration of the particle in the inertial frame of reference can be simplified as:

$$\vec{a}_i = \vec{a}_r + \vec{a}_e + \vec{a}_c \tag{2.2}$$

The physical significance of each term in the formula is as follows:

I.   **Relative Acceleration**

$$\vec{a}_r = \frac{d\vec{v}_r}{dt} = \frac{d^2\vec{r}_r}{dt^2} \tag{2.3}$$

This represents the acceleration of the particle relative to the moving coordinate system. It is equivalent to the acceleration of the particle relative to the inertial frame of reference calculated when the moving coordinate system is static relative to the inertial frame of reference.

## II. Convective Acceleration

$$\vec{a}_e = \frac{d^2\vec{r}_o}{dt^2} + \frac{d\vec{\Omega}}{dt} \times \vec{r}_r + \vec{\Omega} \times (\vec{\Omega} \times \vec{r}_r) \tag{2.4}$$

In the above equation, $\frac{d^2\vec{r}_o}{dt^2}$ represents the convective acceleration of the particle relative to the inertial frame of reference caused by the translational motion of the moving coordinate system, i.e., the motion of the moving coordinate system surrounding its own origin. $\frac{d\vec{\Omega}}{dt} \times \vec{r}_r$ represents the tangential convective acceleration caused by the angular acceleration of the rotating coordinate system. Even if the particle is static relative to the moving coordinate system, the moving coordinate system still rotates with the particle. Therefore, when the moving coordinate system has its own angular acceleration in rotational motion, it will give the particle a tangential convective acceleration. $\vec{\Omega} \times (\vec{\Omega} \times \vec{r}_r)$ is the centripetal convective acceleration created by the angular velocity of the rotating moving coordinate system. If the particle is in a static state relative to the moving coordinate system, it will travel in a circular motion matching the rotation of the moving coordinate system, thus creating centripetal convective acceleration.

## III. Coriolis Acceleration

$$\vec{a}_c = 2\vec{\Omega} \times \vec{v}_r = 2\vec{\Omega} \times \frac{d\vec{r}_r}{dt} \tag{2.5}$$

This equation represents the fact that the relative velocity of the moving coordinate system to the particle is related to the natural rotational angular velocity of the moving coordinate system. Due to the fact that the particle has different locations within the moving coordinate system at different times, its centripetal convective acceleration $\vec{\Omega} \times \vec{r}_r$ changes over time, thus generating acceleration $\vec{\Omega} \times \frac{d\vec{r}_r}{dt}$. Additionally, acceleration $\vec{\Omega} \times \frac{d\vec{r}_r}{dt}$ is also caused due to the rotation of the moving coordinate system changing the direction of relative velocity $\frac{d\vec{r}_r}{dt}$. The sum of these two types of acceleration, given as $2\vec{\Omega} \times \frac{d\vec{r}_r}{dt}$, is the Coriolis acceleration.

In conclusion, Coriolis acceleration is neither relative acceleration nor convective acceleration, but a type of complementary acceleration. It is caused by the following factors: when the moving point is in convective motion, i.e., when the moving frame of reference rotates, the convective rotation causes a constant change in the direction

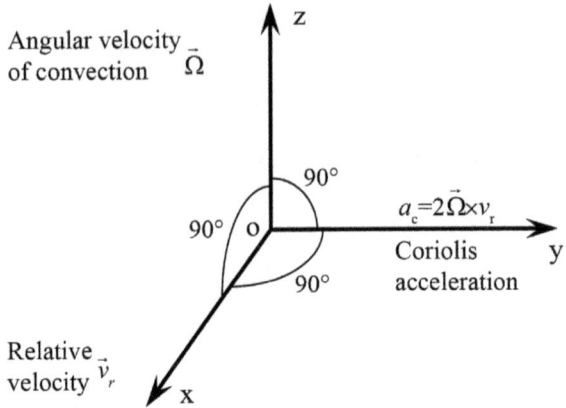

**Fig. 2.2** The direction of Coriolis acceleration

of the relative velocity, while the relative velocity causes a constant change in the convective velocity. The above two changes create a complementary velocity variation rate in the same direction. This complementary velocity variation rate is known as the Coriolis acceleration, and is created under the mutual influence of the relative motion and convective motion. The direction of the Coriolis acceleration is shown in Fig. 2.2. It is perpendicular to the convective angular velocity and relative velocity, and can be determined according to the right-hand precession rule. In a vibratory gyroscope, the Coriolis force enables its sensing structure to detect movement, and is the key to detecting angular velocity.

## 2.2 Working Principles of CVGs

### 2.2.1 Basic Working Principles

The basic working principles of the CVG are as follows. Piezoelectric/electrostatic/electromagnetic force excites the drive mode of the resonator; the Coriolis force effect of the angular velocity of the sensing axis in the gyroscope excites the resonator's sense mode of the resonator; the displacement transducer detects the amplitude of resonator vibrations in sense mode; and finally, the peripheral circuit demodulates the output signal, making it possible to calculate the angular velocity of the sensing axis.

As shown in Figs. 2.3, 2.4 and 2.5, the moment of force acts on the ring/bottom side of the cylindrical resonator, causing "circular-elliptical" flexural vibrations in the resonant ring. When the input signal frequency is consistent with the natural frequency of the gyroscope resonator, the drive mode of the resonator can be excited. When the CVG bears an alternating force, both of its driving moment and response

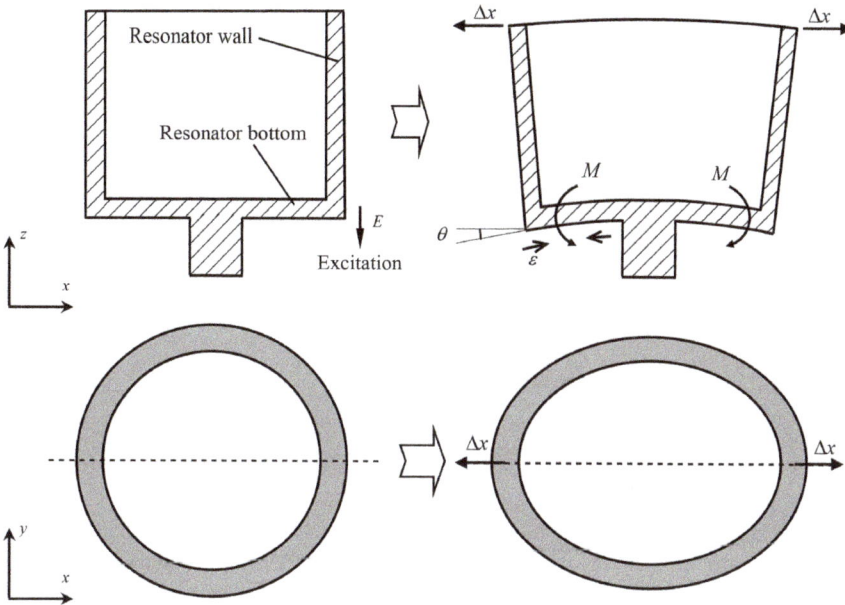

**Fig. 2.3** The excitation signal and drive mode of the CVG

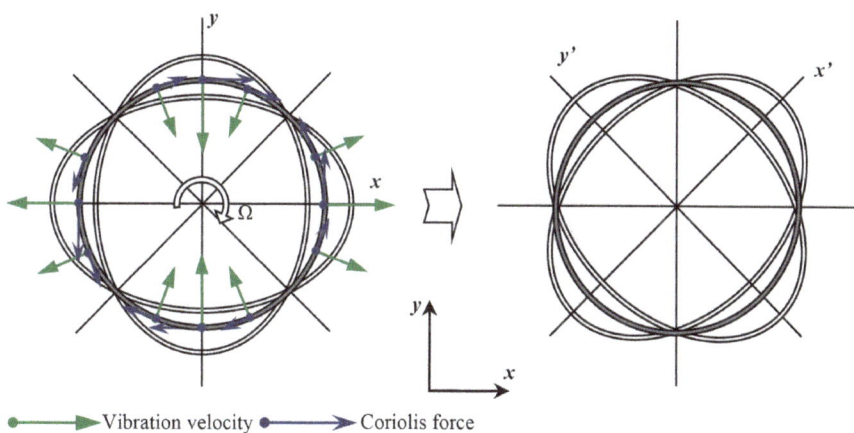

Vibration velocity ●——▶ Coriolis force

**Fig. 2.4** The Coriolis force effect and sense mode of the CVG

displacement are symmetric around the center of the $z$-axis (Fig. 2.3), and its drive mode is a differential vibration mode [2].

As shown in Fig. 2.4a, the drive mode of the CVG resonator consists of a "circular-elliptical" flexural vibration of the resonant ring in the $x$–$y$-axial direction. When angular velocity $\Omega_z$ is input into the sensing axis of the gyroscope, a Coriolis force acts on all points on the resonant ring where there is vibration velocity. The resultant

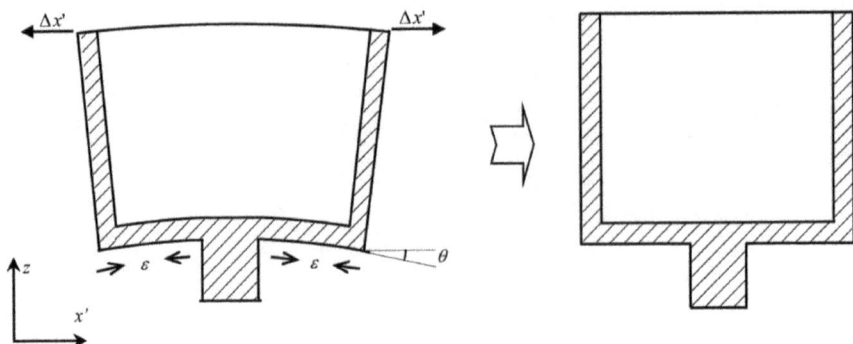

**Fig. 2.5** The Coriolis effect and sense mode of the CVG

Coriolis force at all points on the resonant ring is also a simple harmonic force and excites "circular-elliptical" flexural vibrations in the resonant ring in the $x'-y'$-axial direction. The frequency of the Coriolis force is consistent with the natural frequency of the drive mode of the resonator, suggesting that the Coriolis force can also excite the sense mode of the resonator, as shown in Fig. 2.4b. As can be seen in Fig. 2.4a, b, the Coriolis force within the CVG and its response displacement are both symmetrical around the center of z-axis, and its sense mode is also a differential vibration mode.

The sense mode of the CVG resonator consists of a "circular-elliptical" flexural vibration in the $x'-y'$ direction. As shown in Fig. 2.5, the vibration is detected by the sensor in the resonator, and a detection signal is output. The angular velocity input into the sensing axis of the gyroscope can be calculated by detecting and demodulating the output signal through the control circuit.

### 2.2.2 Excitation and Detection Principles Based on Piezoelectric Electrodes

The piezoelectric effect is widely used in a number of different types of Coriolis vibratory gyroscopes, mainly including gyroscopes with piezoelectric crystals, gyroscopes with piezoelectric electrodes, and gyroscopes with piezoelectric films. Gyroscopes with piezoelectric crystals are divided into two types: gyroscopes with piezoelectric ceramics, and those with quartz crystals. These components not only serve as drive elements and detection elements, but also as the vibratory elements of resonators. A typical representative of the former type is Watson's RSG series, while a typical representative of the latter is the QRS11 quartz micro-gyroscope developed by BEI [3], an American company. Piezoelectric films are usually used in micromachined gyroscopes, and are usually prepared through micro-fabrication process, including sputtering or metalorganic chemical vapor deposition. Piezoelectric electrodes, which can be prepared by advanced manufacturing processes, are flexible in size and shape as well as low in cost, and have been widely applied in the manufacture of sensors.

A CVG uses the piezoelectric effect of piezoelectric electrodes to drive and detect the vibrations of the resonator. It is characterized by high driving force and sensitivity. Moreover, the resonator structure possesses high symmetry as well as reliable driving and detection performance, and gyroscopes of this type are relatively easy to manufacture [4].

As shown in Fig. 2.6, eight piezoelectric electrodes are evenly distributed at the bottom of the resonator. Moreover, these piezoelectric electrodes are mutually identical in terms of their direction of polarization, and are grounded through the resonator's metal structure. The bottom surfaces of each two exciting piezoelectric electrodes on the same diameter are short-circuited to each other and share driving signals, so their vibrations are symmetric around the gyration center of the resonator, exciting the differential vibration mode of the resonator. This short-circuiting also means that each two exciting piezoelectric electrodes on the same diameter jointly output driving signals to produce a greater output of detection signals from the gyroscope. To reduce the signal impact of the electrode lead and solder on the vibrations of the resonator, the lead is welded on the end of each piezoelectric electrode, close to the mounting stem of the resonator. The eight piezoelectric electrodes (four pairs) of the CVG each have different functions. Piezoelectric electrodes 1 and 5 are exciting electrodes for the drive mode; piezoelectric electrodes 3 and 7 are detection electrodes for the drive mode; piezoelectric electrodes 2 and 6 are detection electrodes for the sense mode; and piezoelectric electrodes 4 and 8 are compensating electrodes for the sense mode. The piezoelectric electrodes are therefore not only drive elements, but also detection elements. Their performance determines both the amplitude of the resonator and the intensity of detection signals, thereby greatly influencing the performance of the gyroscope.

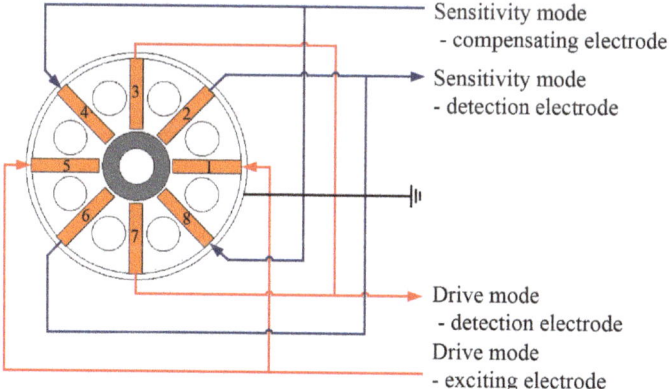

**Fig. 2.6** Distribution of electrodes in a CVG

**Table 2.1** The relationship between stress (strain) subscript and direction

| Subscript | 1 | | 2 | 3 | 4 | | 5 | 6 |
|-----------|---|--|---|---|---|--|---|---|
| Direction | $xx$ | | $yy$ | $zz$ | $yz$ | | $zx$ | $xy$ |
| Meaning | Normal stress (strain) | | | | Shear stress (strain) | | | |

## I. Excitation

The piezoelectricity of piezoelectric materials is related to the interaction between electricity and mechanical behaviors. Since independent variables can be arbitrarily selected for piezoelectric equations, the piezoelectricity of piezoelectric materials usually can be described with four classes of piezoelectric equations [5]. For vibration and waveform analyses, the second class of piezoelectric equation is easier to use. That is, the electric-field strength and strain are selected as independent variables and stress and electric displacement are selected as the dependent variables of the following piezoelectric equations:

$$\begin{cases} T = c^E \cdot S - e^t \cdot E \\ D = e \cdot S + \varepsilon^S \cdot E \end{cases} \tag{2.6}$$

where $S$ and $T$ represent the strain tensor and stress tensor of the piezoelectric element respectively. $D$ and $E$ represent the electric displacement vector and electric-field strength vector respectively. $e$ represents the piezoelectric stress coefficient tensor, $c^E$ represents the elastic stiffness coefficient tensor in a constant electrical field, $\varepsilon^S$ represents the dielectric coefficient tensor under constant strain, and superscript $t$ is a matrix inversion symbol. For the convenience of expression, the relationship between the value subscript of stress and strain tensor elements and the relevant direction are described in Cartesian coordinates in Table 2.1. Similarly, the value subscript of electric-field strength and electric displacement also represent the relevant coordinates.

With a high piezoelectric modulus and good stability, PZT-5 (a piezoceramic material) is widely applied in resonator driving and detection. PZT-5's elastic stiffness coefficient tensor is a symmetric matrix which contains 12 nonzero components, and its piezoelectric stress coefficient tensor contains 5 components. Formula (2.6) can be expressed in components, as shown in Formula (2.7) and Formula (2.8):

$$\begin{bmatrix} T_1 \\ T_2 \\ T_3 \\ T_4 \\ T_5 \\ T_6 \end{bmatrix} = \begin{bmatrix} c_{11} & c_{12} & c_{13} & 0 & 0 & 0 \\ c_{12} & c_{11} & c_{13} & 0 & 0 & 0 \\ c_{13} & c_{13} & c_{33} & 0 & 0 & 0 \\ 0 & 0 & 0 & c_{44} & 0 & 0 \\ 0 & 0 & 0 & 0 & c_{44} & 0 \\ 0 & 0 & 0 & 0 & 0 & (c_{11}-c_{12})/2 \end{bmatrix} \begin{bmatrix} S_1 \\ S_2 \\ S_3 \\ S_4 \\ S_5 \\ S_6 \end{bmatrix} - \begin{bmatrix} 0 & 0 & e_{31} \\ 0 & 0 & e_{31} \\ 0 & 0 & e_{33} \\ 0 & e_{15} & 0 \\ e_{15} & 0 & 0 \\ 0 & 0 & 0 \end{bmatrix} \begin{bmatrix} E_1 \\ E_2 \\ E_3 \end{bmatrix}$$

$$\tag{2.7}$$

$$
\begin{bmatrix} D_1 \\ D_2 \\ D_3 \end{bmatrix} = \begin{bmatrix} 0 & 0 & 0 & 0 & e_{15} & 0 \\ 0 & 0 & 0 & e_{15} & 0 & 0 \\ e_{31} & e_{31} & e_{33} & 0 & 0 & 0 \end{bmatrix} \begin{bmatrix} S_1 \\ S_2 \\ S_3 \\ S_4 \\ S_5 \\ S_6 \end{bmatrix} + \begin{bmatrix} \varepsilon_{11}^S & 0 & 0 \\ 0 & \varepsilon_{11}^S & 0 \\ 0 & 0 & \varepsilon_{33}^S \end{bmatrix} \begin{bmatrix} E_1 \\ E_2 \\ E_3 \end{bmatrix} \qquad (2.8)
$$

Table 2.2 shows the coefficient and component values of each parameter matrix of PZT-5. $\varepsilon_0$ represents the permittivity of vacuum.

Based on the varying relationship of the direction of polarization and the vibration direction of piezoelectric electrodes, piezoelectric electrodes can generate different vibration modes. When the polarization direction is the same as the vibration direction, a longitudinal vibration mode is generated; when the polarization direction is perpendicular to the vibration direction, a lateral vibration mode is generated. The main vibration mode of PZT-5 is the length extension vibration mode, i.e., the lateral vibration mode, as shown in Fig. 2.7.

A flexural piezoelectric actuator can be formed by pasting piezoelectric electrodes to the bottom of the resonator, as shown in Fig. 2.8. The output displacement of the

**Table 2.2** Constant values of PZT-5

| Parameter | Value (Unit) | Parameter | Value (Unit) |
|-----------|--------------|-----------|--------------|
| $c_{11}$ | $14.9 \times 10^{10}$ N/m$^2$ | $c_{33}$ | $13.2 \times 10^{10}$ N/m$^2$ |
| $c_{12}$ | $8.7 \times 10^{10}$ N/m$^2$ | $c_{44}$ | $2.5 \times 10^{10}$ N/m$^2$ |
| $c_{13}$ | $9.1 \times 10^{10}$ N/m$^2$ | | |
| $e_{31}$ | $-4.5$ C/m$^2$ | $\varepsilon_{11}$ | $2100\varepsilon_0$ |
| $e_{33}$ | $17.5$ C/m$^2$ | $\varepsilon_{33}$ | $2400\varepsilon_0$ |
| $e_{15}$ | $15.2$ C/m$^2$ | $\varepsilon_0$ | $8.85 \times 10^{-12}$ F/m |

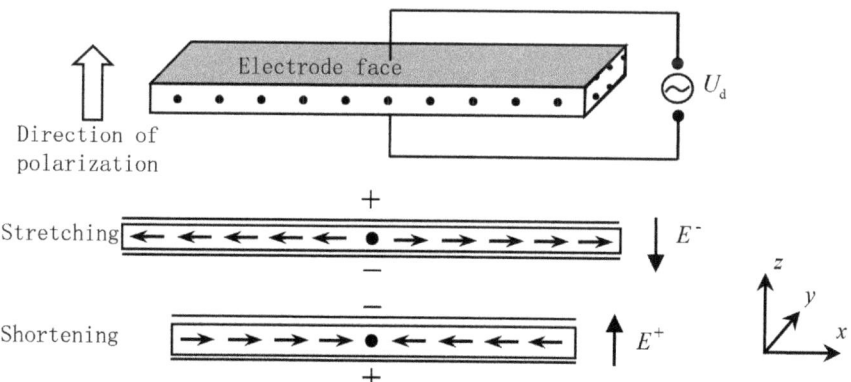

**Fig. 2.7** Schematic of the length extension mode of a polarized piezoelectric electrode along the $z$ axis

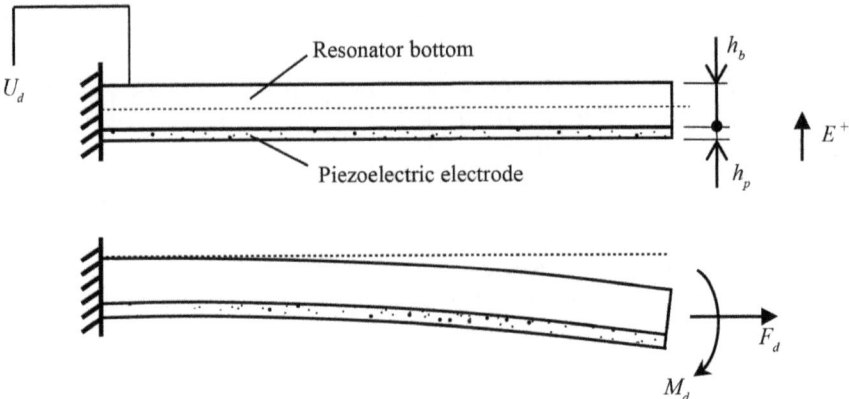

**Fig. 2.8** Structural diagram of a flexural piezoelectric actuator

flexural piezoelectric actuator is perpendicular to the principal plane of piezoelectric electrodes. It is usually a composite structure consisting of a piezoelectric electrode and the bottom of the resonator.

When an AC voltage signal $U_d$ is applied to the piezoelectric electrode at the bottom of the resonator, a $z$-axial electric field will be generated in the beam. Let the upper surface of the piezoelectric electrode (the surface connected to the bottom of the resonator) have a negative potential at a certain moment, while the lower surface has a positive potential at that moment. As shown in Fig. 2.8, the direction of the electric field inside the piezoelectric electrode will be the same as the direction of polarization, and the piezoelectric electrode tends to shorten along its length. Because the piezoelectric electrode is connected to the bottom of the resonator, the bottom of the resonator imposes constraints along the axes of its length and width, but there is no constraint on the axis of its thickness. In other words, electric field $E_3$ along the $z$-axis does not generate normal stress in the $z$-axis or shear stress in the $xy$ plane. An equation of constraint conditions can be established for it as follows:

$$\begin{cases} T_3 = 0, T_6 = 0 \\ S_4 = 0, S_5 = 0 \end{cases} \qquad (2.9)$$

According to piezoelectric Eq. (2.6) of PZT-5, an inner stress equation of the piezoelectric electrode is derived as follows:

$$\begin{cases} T_1 = c_{11}S_1 + c_{12}S_2 + c_{13}S_3 - e_{31}E_3 \\ T_2 = c_{12}S_1 + c_{11}S_2 + c_{13}S_3 - e_{31}E_3 \\ T_3 = c_{13}S_1 + c_{13}S_2 + c_{33}S_3 - e_{33}E_3 \\ T_6 = (c_{11} - c_{12})S_6/2 \end{cases} \qquad (2.10)$$

Further, the constraint of the bottom of the resonator on the piezoelectric electrode is taken into consideration. Because the $y$-axial tensile stiffness and $x$-axial flexural stiffness at the bottom of the resonator are high, the $y$-axial strain of the piezoelectric electrode can be ignored. Equation (2.10) is solved as follows:

$$\begin{cases} S_3 = \frac{e_{33}}{c_{33}}E_3 - \frac{c_{13}}{c_{33}}S_1 \\ T_1 = \left(\frac{c_{13}e_{33}}{c_{33}} - e_{31}\right)\frac{U_d}{h_p} + \left(c_{11} - \frac{c_{13}^2}{c_{33}}\right)S_1 \end{cases} \tag{2.11}$$

As can be seen in Eq. (2.11), the $x$-axial normal stress inside the piezoelectric electrode is related to the applied electric field on the piezoelectric electrode and its $x$-axial strain. Judging from the structural characteristics of the flexural piezoelectric actuator, while the piezoelectric electrode is under the action of driving voltage, the external $x$-axial force acting upon it must be counter-balanced with $x$-axial stress. Owing to the asymmetry of the composite structure of the piezoelectric electrode and the bottom of the resonator, the $x$-axial stress of the piezoelectric electrode also generates a bending moment on the neutral surface of the composite structure, balancing the external bending moment. This can also be understood in the following way: the piezoelectric electrode generates driving force $F_d$ and driving moment $M_d$ under the action of the driving voltage (Fig. 2.8).

Driving force $F_d$ of the piezoelectric electrode can be expressed as:

$$\begin{aligned} F_d &= \int_{h_b/2}^{h_p+h_b/2} T_1(z)b_p dz \\ &= \left(\frac{c_{13}e_{33}}{c_{33}} - e_{31}\right)U_d b_p + \left(c_{11} - \frac{c_{13}^2}{c_{33}}\right)S_1 b_p h_p \end{aligned} \tag{2.12}$$

Driving moment $M_d$ of the piezoelectric electrode can be expressed as:

$$\begin{aligned} M_d &= \int_{h_b/2}^{h_p+h_b/2} T_1(z)b_p z dz \\ &= \int_{h_b/2}^{h_p+h_b/2} \left[\left(\frac{c_{13}e_{33}}{c_{33}} - e_{31}\right)E_3 + \left(c_{11} - \frac{c_{13}^2}{c_{33}}\right)S_1(z)\right]b_p z dz \\ &= \int_{h_b/2}^{h_p+h_b/2} \left[\left(\frac{c_{13}e_{33}}{c_{33}} - e_{31}\right)E_3 + \left(c_{11} - \frac{c_{13}^2}{c_{33}}\right)\left(\frac{2z}{h_p + h_b}\right)S_1\right]b_p z dz \\ &= U_d\left(\frac{c_{13}e_{33}}{c_{33}} - e_{31}\right)b_p\frac{h_b + h_p}{2} + S_1\left(c_{11} - \frac{c_{13}^2}{c_{33}}\right)b_p h_p \frac{3h_b^2 + 6h_b h_p + 4h_p^2}{6(h_p + h_b)} \end{aligned} \tag{2.13}$$

where $b_p$ represents the width of the piezoelectric electrode; $h_p$ represents the thickness of the piezoelectric electrode; $h_b$ represents the thickness of the bottom of the resonator; and $S_1$ refers in particular to the $x$-axial strain on the middle surface of the piezoelectric electrode. Because a piezoelectric electrode is usually thin, the $x$-axial stress gradient on the $y_z$ section inside it can be ignored, while $x$-axial stress $T_1$ on its middle surface can be used in place of the $x$-axial stress function $T_1(z)$ on the $y_z$ section within it. The expression (2.13) for the driving moment can therefore be rewritten as follows:

$$
\begin{aligned}
M_d &= \int_{h_b/2}^{h_p+h_b/2} T_1 b_p z dz \\
&= U_d \left( \frac{c_{13}e_{33}}{c_{33}} - e_{31} \right) b_p \frac{h_b + h_p}{2} \\
&\quad + S_1 \left( c_{11} - \frac{c_{13}^2}{c_{33}} \right) b_p h_p \frac{h_b + h_p}{2}
\end{aligned}
\tag{2.14}
$$

According to Eqs. (2.12) and (2.14), the bottom of the resonator tends to extend and bend under the action of the driving force and driving moment of the piezoelectric electrode. But in fact, within the bottom of the resonator, its radial tensile stiffness $k_F$ and bending stiffness $k_M$ are greatly different from each other and can be respectively expressed as follows:

$$
\begin{cases}
k_F = \frac{F}{\Delta l_x} = \frac{A_b E_b}{l_p} = \frac{h_b b_b E_b}{l_p} \\
k_M = \frac{M}{\Delta u_z} = \frac{2E_b I_b}{l_p^2} = \frac{h_b^3 b_b E_b}{6l_p^2}
\end{cases}
\tag{2.15}
$$

For the action of the driving force and driving moment of the piezoelectric electrode, as shown in Eq. (2.15), the $x$-axial displacement $\Delta l_x$ and $z$-axial $\Delta u_z$ on the outer edge of the bottom of the resonator are respectively:

$$
\begin{cases}
\Delta l_x = \frac{F l_p}{h_b b_b E_b} \\
\Delta u_z = \frac{3 F l_p^2 (h_b + h_p)}{h_b^3 b_b E_b}
\end{cases}
\tag{2.16}
$$

As shown by Eq. (2.16), $\Delta u_z \gg \Delta l_x$, and the tensile deformation of the bottom of the resonator can be ignored when compared with the flexural deformation. It can hence be seen that the piezoelectric electrode and the bottom of the resonator form a flexural piezoelectric actuator, and the entire resonator can be driven to vibrate by applying an AC voltage signal on the piezoelectric electrode. If the frequency of the AC voltage signal happens to be the natural frequency of resonator, the CVG will generate a drive mode.

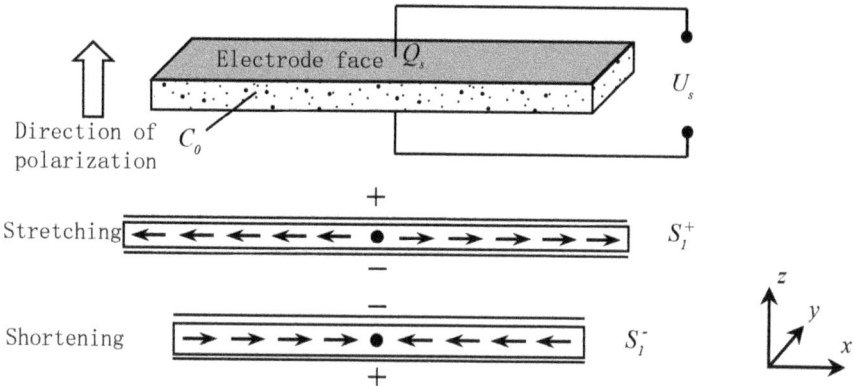

**Fig. 2.9** Z-axial diagram of the piezoelectric effect in a polarized piezoelectric electrode

## II. Detection Principles

According to the results of analyses on the two working modes of the CVG resonator, there is a "circular-elliptical" flexural vibration mode on its shell wall and a flexural vibration mode at its bottom. It can thus be seen that in the piezoelectric electrode has flexural vibrations with the bottom of the resonator in the CVG's working mode.

According to the relationship between the direction of polarization of the piezo-electric electrode and bound charges, when the z-axial polarized piezoelectric electrode deforms with the resonator, the positive and negative charge centers inside the material will be relatively displaced, polarizing the electrode and creating opposite charges on the two electrode surfaces relative to the material. Moreover, the charge density will be directly proportional to the strain in the piezoelectric electrode, as shown in Fig. 2.9.

According to the piezoelectric effect principle acting on a z-axial polarized piezo-electric electrode, a flexural piezoelectric sensor can be made by attaching a piezo-electric electrode to the bottom of the resonator, as shown in Fig. 2.10. This flexural piezoelectric sensor can be used to detect the flexural deformation on the principal plane of the piezoelectric electrode, and its structure is the same as the flexural piezoelectric actuator.

According to Euler–Bernoulli beam theory, after the piezoelectric electrode gener-ates a deflection at the value of $u$ off of the bottom of the resonator, the strain in the piezoelectric electrode is mainly the x-axial normal strain. The x-axial normal strain on its neutral surface is related to the deflection curve at the bottom of the resonator. Because a piezoelectric electrode is usually thin, the z-axial stress gradient on the $y_z$ section inside it can be ignored, while the x-axial strain $S_1(x)$ on its middle surface can be used in place of the function $S_1(x, z)$ to calculate z-axial changes of the x-axial strain on its inner $yz$ section. According to the piezoelectric equation of the piezo-electric electrode, in the absence of an applied electrical field, the x-axial strain $S_1(x)$ tends to generate electrical displacement $D_3$ on the $z$ surface of the piezoelectric electrode. The expression for this is as follows:

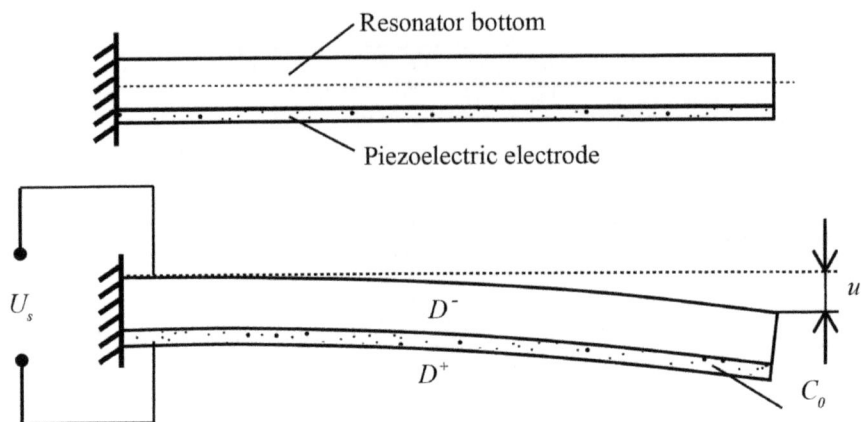

**Fig. 2.10**  Structural diagram of a flexural piezoelectric sensor

$$D_3(x) = e_{31} S_1(x) \tag{2.17}$$

Detection charge $Q_s$ on the upper and lower surfaces of the piezoelectric electrode can be expressed as:

$$Q_s = \int_{A_z} D_3 dA = \int_0^{l_p} e_{31} S_1(x) b_p dx \tag{2.18}$$

A CVG has small resonance amplitude, which is negligible compared with the size of the piezoelectric electrode. Therefore, the static capacitance $C_0$ between the $z$-axial surfaces of the piezoelectric electrode can be approximated as the capacitance between parallel plates. Its expression is as follows:

$$C_0 = \varepsilon_{33} \frac{b_p l_p}{h_p} \tag{2.19}$$

The output detection voltage $U_s$ of the piezoelectric electrode can be calculated according to Eqs. (2.18) and (2.19). The result is as follows:

$$U_s = \frac{Q_s}{C_0} = \frac{e_{31} h_p S_1}{\varepsilon_{33}} = \frac{e_{31} h_p \int_0^{l_p} S_1(x) dx}{\varepsilon_{33} l_p} \tag{2.20}$$

As can be seen, the piezoelectric electrode and the bottom of the resonator form a flexural piezoelectric sensor, which can be used to detect the flexural deformation of the resonator bottom and output detection voltage. In the CVG's two working modes, whether they are in drive mode or sense mode, there is a fixed relative relationship among the vibrations at all points on the resonator. The flexural piezoelectric sensor can therefore be used to detect the vibrations at all points on the resonator.

## 2.3  Typical CVG Structures

A standard CVG resonator consists of a cylindrical shell and piezoelectric electrodes. This structure is characterized by good symmetry, ease of manufacturing, low technical requirements for advanced processing equipment, easily controlled costs, and high suitability for mass production.

As shown in Fig. 2.11, the resonator is a revolving body and is symmetric around the sensing axis of its angular velocity (the center axis of the revolving body). The resonator structure is composed of a shell wall, a bottom, and a support column. These eight piezoelectric electrodes are evenly distributed along the basal plane of the resonator and are used to excite and detect the vibrations of the resonator in different modes. The shell wall of the resonator is its major inertial mass component and is used to generate a gyroscopic effect. Its high degree of stiffness and mass determine the main vibration characteristics of the resonator; and its support column is used to connect the resonator with the packaging shell. The lower surfaces of the eight piezoelectric electrodes are connected to the base through a lead wire and ultimately to the peripheral measurement and control circuit. The sealed space/vacuum formed by the seal cover and mounting base ensures the stable vibration of the resonator while working. Figure 2.12 shows a packaged CGV resonator.

The following are the main characteristics of a CVG:

(1)   A CVG has a fully symmetrical structure, which results in high sensitivity. The specially designed cylindrical shell wall and planar bottom separate the driving and detection components from the inertial mass component, reducing the impact of the mass of the driving and detection electrodes on the dynamic characteristics of the resonator.

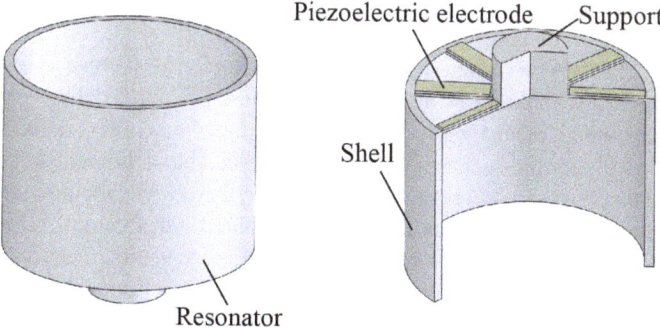

**Fig. 2.11**  Typical structure of CVG resonators

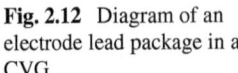

**Fig. 2.12** Diagram of an electrode lead package in a CVG

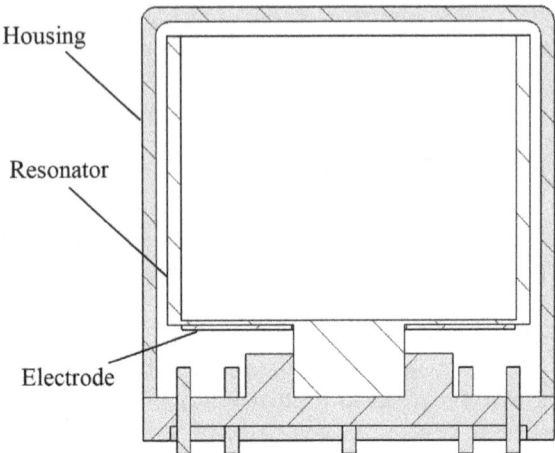

Housing

Resonator

Electrode

(2)  There are many materials available for selection, and these gyroscopes can be manufactured in different ways. A gyroscope resonator made of a high-performance alloy material is featured by high sensitivity, low processing difficulty, low costs, and high suitability for mass production.

(3)  Piezoelectric electrodes are adopted as the gyroscope's driving and detection components, providing it with significant driving force and high sensitivity. Moreover, the high symmetry of the resonator structure and the reliability of the driving and detection performance are ensured, while the manufacturing difficulty of the gyroscope has been reduced.

## 2.4  Derivative CVG Structures

Although a typical CVG is characterized by the advantages of a simple structure and ease of manufacturing, it also has obvious disadvantages, such as a low Q value and poor temperature stability. Some researchers have therefore developed improved CVG structures which provide some optimization effects.

Figure 2.13a shows a resonator of varying wall thickness. It consists of upper and lower cylindrical shells of differing wall thicknesses. The upper cylindrical shell has a thicker wall, which means improved stiffness, reduced machining deformation and higher quality of vibrations. The lower cylindrical shell has a thinner wall, which can increase the response displacement of the resonator, thereby enhancing the sensitivity of the gyroscope. The round holes uniformly distributed in the bottom of the resonator are used to cut off vibrational coupling among the piezoelectric electrodes in order to avoid interference with both modes. Compared with traditional CVG structures, the varying wall thickness greatly increases the Q value of the resonator [6].

Figure 2.13b shows a gyroscope with a beam structure. The resonant ring is supported by an axial beam, while a piezoelectric beam is stuck to the support beam

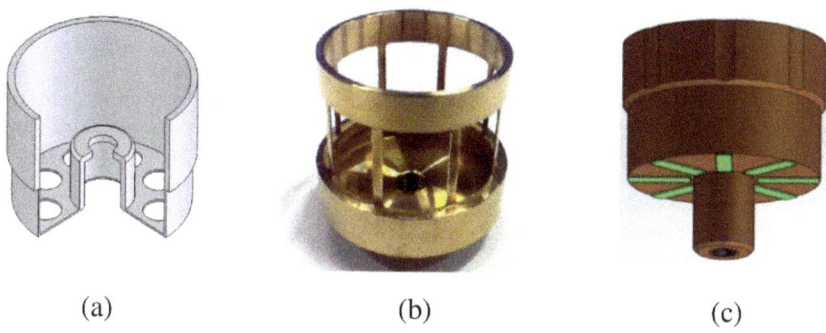

**Fig. 2.13** Diagram of some derivative CVG structures **a** A resonator of varying wall thickness, **b** a resonator with a beam structure, and **c** a multilobe resonator

[7]. The resonant ring is precisely machined using a lathe, while the sidewall is made into a beam structure using EDM. This design can simplify the machining process and lower production costs, because it enables the resonator to ensure high precision in the resonant ring.

Figure 2.13c shows a shell vibratory gyroscope with a circumferential evolution structure [8]. This structure is characterized by the presence of changing protrusions and dimples on the resonant ring, with the mass concentrated in the direction of the shell's vibration. Moreover, a periodic temperature stress distribution is formed, making it easy to maintain a stable vibration mode even when the temperature changes.

In short, with the continuous improvement of modern machining processes, there is a rapid development for CVGs. The optimization and improvement of their structures will serve to increase the Q value and sensitivity of the resonator and simplify its machining process.

# References

1. Shkel, A. M. (2006). Type I and Type II micromachined vibratory gyroscopes. In *Proceedings of the position, location, and navigation symposium*, San Diego, California, USA, 25–27 Apr 2006.
2. Tao, Y. (2011). *Research on the key technologies of cup-shaped wave gyros*. Changsha: National University of Defense Technology.
3. Nordall, B. D. (1994). Quartz fork technology may be replaced in gyros. *Aviation Week & Space Technology, 140*(17), 4.
4. Tan, P. (2005). Research on the active control of bonded piezoelectric ceramic actuators. *Journal of Nanjing University of Science and Technology, 29*(6), 2.
5. Zhang, F. (1981). *Gyros with piezoelectric crystals*. Beijing National Defense Industry Press.
6. Chikovani, V., Yatsenko, Y., Barabashov, A., Marusyk, P., Umakhanov, E., & Taturin, V. (2009). Improved accuracy metallic resonator CVG. *Aerospace and Electronic Systems Magazine, 24*(5), 40–43.

7. Tao, Y., Wu, X., Xiao, D., Wu, Y., Cui, H., Xi, X., & Zhu, B. (2011). Design, analysis, and experiment of a novel ring vibratory gyroscope. *Sensors and Actuators a-Physical, 168*(2), 286–299.
8. Xinxin, X., Yulie, W., Xiang, X., Hanhui, H. (2015). Characterization and performance tests of a novel petal-shaped vibratory shell gyroscope. In *Proceedings of the 6th international conference of Asian society for precision engineering and nanotechnology*, Harbin, China, 15–20 August, 2015.

# Chapter 3
# Theoretical Analysis and Modeling of CVGs

## 3.1 Basic Mathematical Models of Resonators

The resonator of a CVG has a typical thin-walled structure. Under normal conditions, the theory of plates and shells can be used to study the vibration characteristics of a thin-walled structure. The theory of thin plates is grounded in the famous Kirchhoff hypothesis, while the theory of thin shells is based on the famous Love hypothesis [1–3].

The Kirchhoff hypothesis of thin plates includes the following: (1) A plate material is an ideal elastomer that is both homogeneous and continuous; (2) Displacement and deformation are very limited, the thin plate thickness is very small compared with the minimum contour dimension, the maximum plate deflection is very small compared with its thickness, and both strain and rotation angle are much less than 1; (3) Because the plate thickness is very small, the parallel layers in the middle plane of the plate do not squeeze each other after deformation, and the (lateral) stress perpendicular to the plate direction can be ignored; (4) Any straight line perpendicular to the middle plane remains a straight line after the plate suffers from flexural deformation, and is perpendicular to the middle plane after deflection. This means that shear deformation is ignored, i.e., the middle plane is not stretched; (5) The moment of inertia caused by bending is ignored, while consideration is given to the inertia caused by the lateral displacement of the plate and the rotation of the segment.

Love hypothesis of thin shells: (1) A shell material is an ideal elastomer that is homogeneous and continuous; (2) The thickness of a shell is thinner than its other dimensions (e.g., the minimum curvature radius of the middle plane of the shell); (3) Strain and displacement are sufficiently small that the second-order and higher-order quantities in the strain–displacement relation can be ignored; (4) The radial normal stress is negligible compared to other normal stresses; (5) The normal of the centerline of each segment remains in a straight line after deformation and perpendicular to the centerline, and the length of the centerline remains unchanged.

According to the structural characteristics of the CVG resonator and the Kirchhoff–Love hypothesis, the bottom of the resonator can be treated as equivalent to a

© National Defense Industry Press 2021
X. Wu et al., *Cylindrical Vibratory Gyroscope*, Springer Tracts in Mechanical Engineering,
https://doi.org/10.1007/978-981-16-2726-2_3

ring-shaped thin plate model (the inner diameter of the ring is the diameter of the mounting stem) while the shell wall of the resonator can be equivalent to a cylindrical shell model, as shown in Fig. 3.1, where $R_0$ refers to the inner diameter of the ring-shaped thin plate. $h_b$ refers to the thickness of the ring-shaped thin plate, $R$ refers to the middle plane radius of the cylindrical shell, $H$ refers to the height of the cylindrical shell, and $t$ refers to the thickness of the cylindrical shell. A cylindrical coordinate system $(r, \theta, z)$ is used to describe the position of each point on the resonator. Moreover, a local coordinate system $(w, v, u)$ is established at each point on the resonator to describe axial $(z)$ displacement component $u$, circumferential $(\theta)$ displacement component $v$, and radial $(r)$ displacement component $w$ of each point on the resonator.

On a ring-shaped thin plate, its natural vibrations can be divided into out-of-plane vibrations (flexural vibrations) and in-plane vibrations. The former means that the vibration direction is perpendicular to the thin plate, while the latter means that the vibration direction is in the plane of the thin plate. An elastic mechanical analysis can then be made on it according to the Kirchhoff hypothesis. The following is the differential equation for the undamped motion of a ring-shaped thin plate in the polar coordinate system:

$$\frac{\partial^2 u}{\partial t^2} + \frac{D}{\rho h_b} \nabla^2 \nabla^2 u = 0 \tag{3.1}$$

**Fig. 3.1** The plate-shell model and geometric coordinate system of the CVG resonator

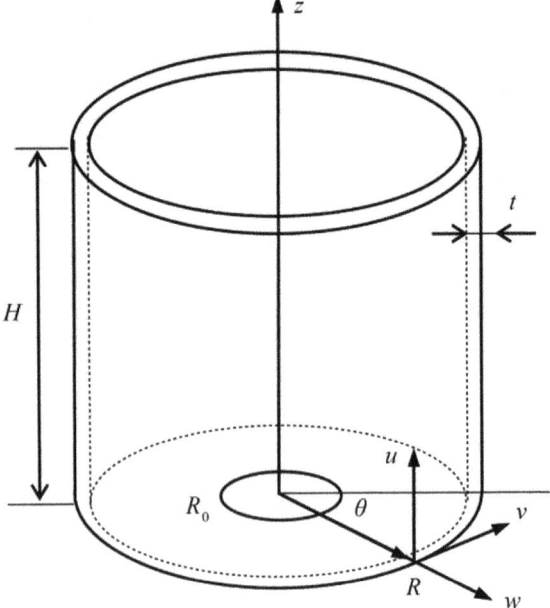

where $\nabla^2 = \frac{\partial^2}{\partial r^2} + \frac{1}{r}\frac{\partial}{\partial r} + \frac{1}{r^2}\frac{\partial^2}{\partial \theta^2}$ is a Laplacian operator, $D = \frac{Eh_b^3}{12(1-\mu^2)}$ represents the flexural stiffness of the plate, $\mu$ represents the Poisson's ratio, and $\rho$ represents the material density.

The solution of Eq. (3.1) is a superposition of numerous simple harmonic vibrations. Its general solution is as follows:

$$u(r, \theta, t) = \sum_{n=1}^{\infty} u_n(r, \theta, t) = \sum_{n=1}^{\infty} u_n(r, \theta)(A_n \sin \omega_n t + B_n \cos \omega_n t)$$

$$= \sum_{n=1}^{\infty} U_n(r)(S_n \sin n\theta + C_n \cos n\theta)(A_n \sin \omega_n t + B_n \cos \omega_n t) \quad (3.2)$$

where $U_n(r)$ is a function for the mode shape of each order, and $\omega_n$ represents the natural frequency of relevant simple harmonic vibrations.

Equation (3.2) is substituted into Eq. (3.1), producing the general solution of $U_n(r)$ as follows:

$$U_n(r) = a_n J_n(kr) + b_n H_n(kr) + c_n I_n(kr) + d_n K_n(kr) \quad (3.3)$$

where $J_n(kr)$ is the Bessel function of the first kind of the $n$th order, $H_n(kr)$ is the Bessel function of the second kind of the $n$th order, $I_n(kr)$ is the modified Bessel function of the first kind of the $n$th order, $K_n(kr)$ is the modified Bessel function of the second kind of the $n$th order, and $a_n$, $b_n$, $c_n$ and $d_n$ are undetermined parameters related to the geometric boundary conditions and mechanical boundary conditions.

The bottom of the CVG resonator is a ring-shaped thin plate. The inner and outer edges of the ring-shaped thin plate are two concentric circumferences; the combined boundary condition is that the inner circle provides fixed support (the mounting stem is rigid) while the outer circle offers an elastic constraint (the shell wall of the resonator has an elastic vibration component). The general solution of its differential equation of motion can also be expressed in Eq. (3.3).

In the resonator, the center of the ring-shaped thin plate coincides with the center of the polar coordinates, and there is finite displacement at the center. Terms $H_n(kr)$ and $K_n(kr)$ in Eq. (3.3) must be discarded. On the other hand, the boundary conditions of the ring-shaped thin plate are symmetrical to any straight line, and the circumferential starting point of the mode shape function can be set to any position, so the term in Eq. (3.2) can also be discarded. Thus, the natural vibration equation of the ring-shaped thin plate at the bottom of the resonator in a mode of a certain order can be expressed as follows:

$$u_n(r, \theta, t) = [a_n S_n J_n(kr) + c_n S_n I_n(kr)] \cos n\theta \cos \omega_n t \quad (3.4)$$

During the vibrations of the ring-shaped thin plate, the displacement of one or more circles concentric with the boundary is zero, and these concentric circles are known

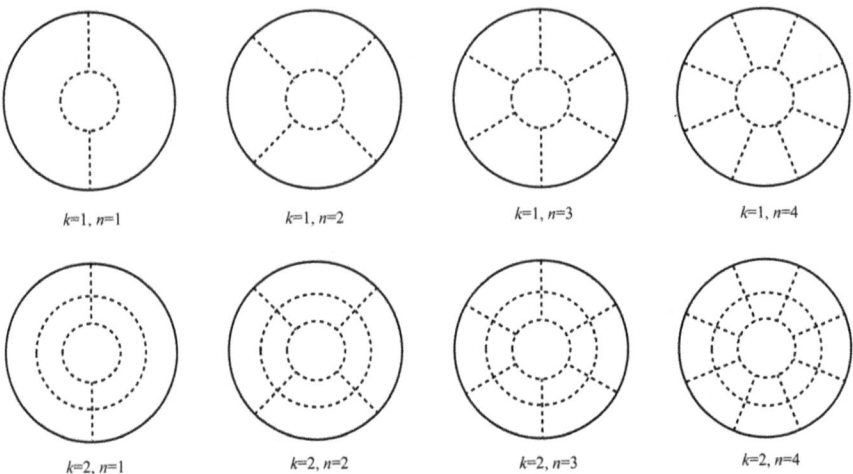

$k=1, n=1$          $k=1, n=2$          $k=1, n=3$          $k=1, n=4$

$k=2, n=1$          $k=2, n=2$          $k=2, n=3$          $k=2, n=4$

**Fig. 3.2** Diagram of the low-order flexural mode shapes of the ring-shaped thin plate

as pitch circles. Similarly, there are also one or more diameters whose displacement remains zero, and such a diameter is known as a pitch diameter. In Eq. (3.4), $k$ represents the number of pitch circles and $n$ represents the number of pitch diameters. Figure 3.2 shows the low-order flexural mode shapes of the ring-shaped thin plate with fixed inner circles as well as the distribution of pitch circles and pitch diameters.

The vibrations of a cylindrical shell are very complicated, and its displacements exist in three directions. To date, a universally recognized and unified theory has yet to be produced on the topic. Commonly used theories in the field include the Love-Timoshenko theory, Donnell-Mushtari theory, Vlasov theory, and Epstein theory. For the small-displacement vibrations of a classic cylindrical shell, the Love hypothesis is mostly used to perform elastic mechanical analysis. Let the displacement vectors of the cylindrical shell in three directions be:

$$\{u_i\} = [\, u \ v \ w \,]^T \tag{3.5}$$

The differential equation of the cylindrical shell is as follows:

$$[\Phi_{D-M}]\{u_i\} = [\, u \ v \ w \,]^T \tag{3.6}$$

where $[\Phi_{D-M}]$ is a Donnell-Mushtari matrix operator. Its expression is as follows:

$$[\Phi_{D-M}] = \begin{bmatrix} \dfrac{\partial^2}{\partial s^2} + \dfrac{1-\mu}{2}\dfrac{\partial^2}{\partial \theta^2} - \\ \rho\dfrac{(1-\mu^4)R^2}{E}\dfrac{\partial^2}{\partial t^2} & \dfrac{1+\mu}{2}\dfrac{\partial^2}{\partial s\partial\theta} & \mu\dfrac{\partial}{\partial s} \\[2em] \dfrac{1+\mu}{2}\dfrac{\partial^2}{\partial s\partial\theta} & \dfrac{\partial^2}{\partial\theta^2} + \dfrac{1-\mu}{2}\dfrac{\partial^2}{\partial s^2} - \\ \rho\dfrac{(1-\mu^4)R^2}{E}\dfrac{\partial^2}{\partial t^2} & \dfrac{\partial}{\partial\theta} \\[2em] \mu\dfrac{\partial}{\partial s} & \dfrac{\partial}{\partial\theta} & 1 + \dfrac{h^2}{12R^2}\nabla^2\nabla^2 + \\ \rho\dfrac{(1-\mu^4)R^2}{E}\dfrac{\partial^2}{\partial t^2} \end{bmatrix}$$

$$(3.7)$$

where $s = z/R$ represent an axial nondimensional coordinate, and $\nabla^2 = \dfrac{\partial^2}{\partial s^2} + \dfrac{\partial^2}{\partial \theta^2}$ is a Laplace operator.

For the vibrations of an infinite cylindrical shell, its displacement function can usually be established as follows [4]:

$$\begin{cases} u = A\cos\lambda s\cos n\theta\cos\omega_n t \\ v = B\sin\lambda s\sin n\theta\cos\omega_n t \\ w = C\sin\lambda s\cos n\theta\cos\omega_n t \end{cases} \qquad (3.8)$$

where $A$, $B$, $C$, and $\lambda$ are undetermined parameters related to the geometric boundary conditions and mechanical boundary conditions; $\omega_n$ refers to the natural frequency of simple harmonic vibrations of each order. It is assumed in the equation that time can be separated from other quantities and that the vibration period and phase are identical at each point on the cylindrical shell.

During the vibrations of a cylindrical shell, the displacement of one or more circles concentric with the boundary remains zero, and these circles are known as pitch circles. Similarly, there are also one or more lines whose displacement remains zero, and such a line is known as a pitch line. In Eq. (3.8), $\lambda$ determines the position of a pitch circle and $n$ represents the number of pitch lines.

The shell wall of the resonator is a finite cylindrical shell, supported by a ring-shaped plate. For the shell wall of a resonator, the bottom of the resonator, which has a high degree of stiffness in its plane, can prevent the shell wall from being stretched, compressed, and shear-cut in its plane, but cannot resist against any deformation perpendicular to its plane. Under these circumstances, the cylindrical shell is a closed cylindrical shell with a free end and a shear film on the other end.

According to Donnell-Mushtari theory, a finite cylindrical shell with a shear film on either fixed ends or free ends (e.g., a cylindrical shell resonator) can have a displacement function assumed to be the following:

$$\begin{cases} u = A_m R'_m(z) \cos n\theta \cos \omega_n t \\ v = B_m R_m(z) \sin n\theta \cos \omega_n t \\ w = C_m R_m(z) \cos n\theta \cos \omega_n t \end{cases} \tag{3.9}$$

where $R_m(z)$ is the displacement distribution characteristic function of the generatrix of the cylindrical shell. It is not only a displacement distribution of a beam with the same boundary conditions as the generatrix on both ends, but is also the mth-order mode shape function of the natural vibrations of the beam. $R'_m(z)$ is the first derivative of $R_m(z)$ with respect to $z$. $A_m$, $B_m$, and $C_m$ are amplitude coefficients corresponding to Donnell-Mushtari shell theory. All modes acquired under the circumstances are generally coupled in three directions. Similarly, there are $2n$ pitch lines on the shell wall of the resonator, corresponding to node position in its circumferential mode. There may also be pitch circles in the axial direction of the shell wall, corresponding to the nodes in the mth-order mode shape function of its generatrix mode. Figure 3.3 shows the low-order flexural mode shapes of the cylindrical shell and the distribution of the corresponding pitch diameters [5].

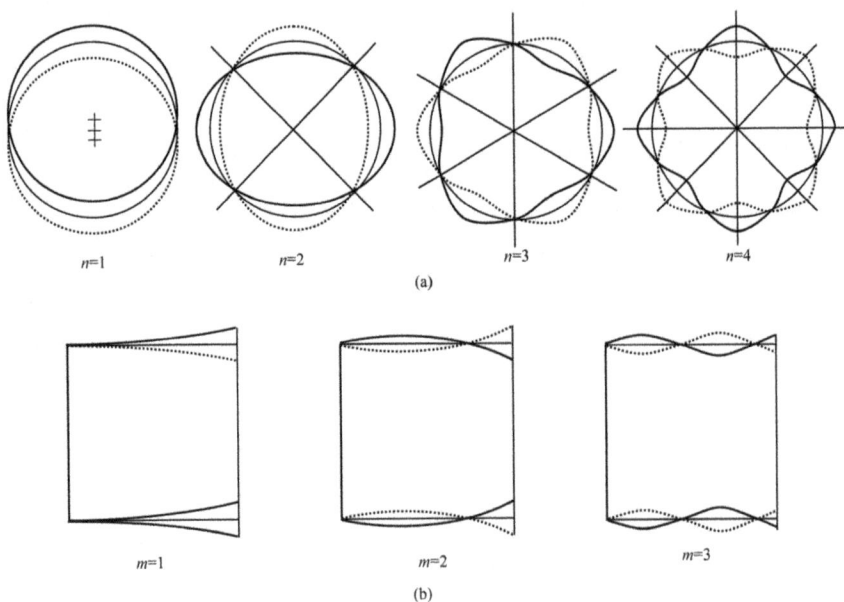

**Fig. 3.3** Diagram of the low-order flexural mode shapes of the cylindrical shell **a** mode shape along the circumference; **b** mode shape along the generatrix.

## 3.2 Analysis of Resonator Stiffness

The cylindrical shell of the CVG resonator is the gyroscope's main inertial mass component, and is used to generate gyroscopic effects. It has high effective stiffness and effective mass, and these determine the main vibration characteristics of the resonator. In the resonator's working modes, the cylindrical shell primarily shows radial, tangential, and axial vibrations. The shell wall of the resonator can be discretized into a number of rings of finite height in the axial direction. In the present section, we will discuss the relationship between the radial, tangential, and axial displacements of the rings and their stiffness under the action of symmetrical forces to prepare for the building of a discrete mechanical model depicting the entire resonator. All analyses of the rings in this section are based on Bernoulli–Euler beam theory, ignoring the lateral shear deformation and warpage [6].

### 3.2.1 Analysis of Radial Stiffness

To reveal the relationship between the displacements and strain energy of the resonant rings in the CVG, a cylindrical coordinate system $(r, \theta, z)$ is established along with a local coordinate system $(u, v, w)$ for the displacement at each point on the resonant rings, as shown in Fig. 3.4. According to the structural characteristics of the CVG resonator, the wall thickness of the resonator (which is much smaller than its radius) is part of the shell, so the vibrations of the resonator can be analyzed using the structural theory of shells.

The deformation of each ring in its own plane from driving forces must be studied to calculate the static displacement of micro-high rings generated by driving forces. We can take a segment of a ring, denoted by $AB$, to study its flexural deformation, as shown in Fig. 3.5.

**Fig. 3.4** Geometric coordinate system of rings

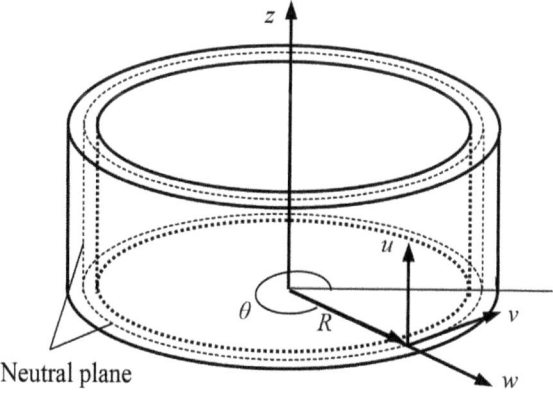

**Fig. 3.5** Flexural
deformation of a resonant
ring

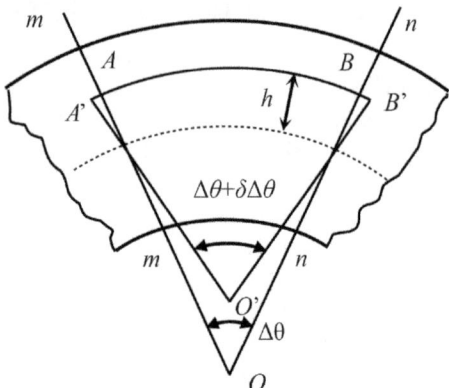

Deduction is based on the precondition that the centerline of the shell wall is not stretched. An arc section, $AB$, is taken from a certain part of the resonant ring with specific thickness, and the length of the undeformed arc is $(R + h)\Delta\theta$. When the wall is bent, the center of the curve moves from point $O$ to point $O'$, and the curve changes from $AB$ to $A'B'$. Its segmental rotation angle is $\delta\Delta\theta$, and the arc length in its deformation state is $(R' + h)\delta\Delta\theta$, so the length of curve $AB$ should be converted into:

$$\varepsilon = \frac{(R' + h)(\Delta\theta + \delta\Delta\theta) - (R + h)\Delta\theta}{(R + h)\Delta\theta} \tag{3.10}$$

According to the Love hypothesis, we know that the centerline is not stretched, so,

$$R'(\Delta\theta + \delta\Delta\theta) = R\Delta\theta \tag{3.11}$$

For a resonant ring, the radius of its centerline is much larger than the wall thickness, so $R \gg h$. Equation (3.11) is substituted into Eq. (3.10), deriving the following equation:

$$\varepsilon = \frac{h\delta\Delta\theta}{R\Delta\theta} = h\left(\frac{1}{R'} - \frac{1}{R}\right) = h\Delta\chi \tag{3.12}$$

where $\chi$ represents the curvature of the centerline. The moment resulting from the stress on segment $mm$ is expressed as follows:

$$M = \int h\sigma ds = \int h^2 E\Delta\chi ds = EI_z\Delta\chi \tag{3.13}$$

**Fig. 3.6** Diagram of the micro-segment displacement of the resonant ring

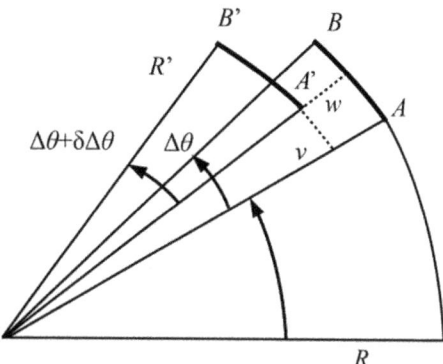

where $E$ represents the Young's modulus of the resonator material, $\Delta\chi$ represents the variation quantity of the ring curvature, and $I_Z$ represents the moment of inertia of the ring segment around its centerline parallel to the z-axis.

After the micro-segment of the ring is bent, radial displacement $w$ and tangential displacement $v$ will be generated in its local coordinate system as shown in Fig. 3.6. The middle plane of the ring is not stretched, so,

$$(R - w)(\Delta\theta + \delta\Delta\theta) = R\Delta\theta \tag{3.14}$$

The angular variation of segment $AB$ is related to the difference in distance between point $A$ and point $B$ in the tangential direction, therefore:

$$\delta\Delta\theta = \frac{\partial v}{\partial\theta}\frac{1}{R}\Delta\theta \tag{3.15}$$

Equation (3.15) is substituted into Eq. (3.14). When the angle of segment $AB$ and radial displacement $w$ approach zero, the following equation can be derived:

$$w = \frac{\partial v}{\partial\theta} \tag{3.16}$$

The rotation angle $\varphi$ of the tangent line $AB$ is determined by its tangential displacement $v$ and radial displacement $w$. The expression for this is as follows:

$$\varphi = \frac{v}{R} + \frac{\partial w}{\partial\theta}\Delta\theta/R\Delta\theta = \frac{1}{R}\left(v + \frac{\partial w}{\partial\theta}\right) \tag{3.17}$$

This shows that the curvature variation $\Delta\chi$ of the tangent line $AB$ is related to its elastic displacement, which can be written as follows:

$$\Delta\chi = \frac{\partial\varphi}{R\partial\theta} = \frac{1}{R^2}\left(w + \frac{\partial^2 w}{\partial\theta^2}\right) \tag{3.18}$$

Equation (3.18) is substituted into Expression (3.13) for internal moment $M$ of the ring, with the following statement able to be derived:

$$M = -\frac{E I_z}{R^2}\left(w + \frac{\partial^2 w}{\partial \theta^2}\right) \tag{3.19}$$

Based on Eqs. (3.18) and (3.19), an expression for the flexural strain energy of the ring can be derived as follows:

$$V = \frac{1}{2}\int M \Delta\chi R d\theta = \frac{1}{2}\frac{E I_z}{R^3}\int\left(w + \frac{\partial^2 w}{\partial \theta^2}\right)^2 d\theta \tag{3.20}$$

According to the structural characteristics of the ring, the solution of Eq. (3.16) can be set as:

$$\begin{cases} v = R\sum\limits_{n=1}^{\infty}(a_n\cos n\theta + b_n\sin n\theta) \\ w = R\sum\limits_{n=1}^{\infty}n(-a_n\sin n\theta + b_n\cos n\theta) \end{cases} \tag{3.21}$$

where $a_n$ and $b_n$ are undefined constants and can be calculated according to the load conditions.

According to our discussion, when $n = 1$, the displacement at each point on the centerline of the ring can be expressed as follows:

$$\begin{cases} v_1 = R(a_n\cos\theta + b_n\sin\theta) \\ w_1 = R(-a_n\sin\theta + b_n\cos\theta) \end{cases} \tag{3.22}$$

The displacement described in Eq. (3.22) is a rigid displacement of the ring. It does not strain the ring, so it doesn't change the strain energy of the ring. Therefore, Eq. (3.22) can be modified as:

$$\begin{cases} v = R\sum\limits_{n=2}^{\infty}(a_n\cos n\theta + b_n\sin n\theta) \\ w = R\sum\limits_{n=2}^{\infty}n(-a_n\sin n\theta + b_n\cos n\theta) \end{cases} \tag{3.23}$$

Equation (3.23) is substituted into Eq. (3.20), with the following equation being able to be derived:

$$V = \frac{1}{2}\frac{E I_z}{R^3}\int\left(w + \frac{\partial^2 w}{\partial \theta^2}\right)^2 d\theta$$

$$= \frac{1}{2} \frac{E I_z}{R} \sum_{n=2}^{\infty} \int_0^{2\pi} \left[ n(-a_n \sin n\theta + b_n \cos n\theta) + n^3 (a_n \sin n\theta - b_n \cos n\theta) \right]^2 d\theta$$

$$= \frac{1}{2} \frac{E I_z}{R} \sum_{n=2}^{\infty} \left( n^3 - n \right)^2 \int_0^{2\pi} (a_n \sin n\theta - b_n \cos n\theta)^2 d\theta$$

$$= \frac{\pi}{2} \frac{E I_z}{R} \sum_{n=2}^{\infty} \left( n^3 - n \right)^2 \left( a_n^2 + b_n^2 \right) \tag{3.24}$$

The internal strain energy of the ring is related to the displacement and deformation of the ring, while the displacement and deformation are related to specific load conditions.

The distribution of the ring's displacement caused by external forces can be analyzed according to the relationship between the strain energy of the ring and its displacement. As indicated by the working principles of the CVG, the resonator is driven by a pair of piezoelectric electrodes. It can therefore be assumed that a pair of forces of the same magnitude but in opposite directions act on two points on the same diameter on the centerline of the ring, as shown in Fig. 3.7.

According to the constitutive equations in elasticity mechanics, driving force $F$ works nowhere except at its point of application, i.e., where $\theta = 0$ and $\theta = \pi$, and thereby deforms the ring. Because the driving force only works in the radial direction of the ring, according to Expression (3.23) for radial displacement, all terms containing undetermined coefficient $a$ are equal to zero where $\theta = 0$ and $\theta = \pi$ so the expression for deformation only has terms with a coefficient of $b_n$, therefore:

$$\begin{cases} v = R \sum_{n=2}^{\infty} b_n \sin n\theta \\ w = R \sum_{n=2}^{\infty} n b_n \cos n\theta \end{cases} \tag{3.25}$$

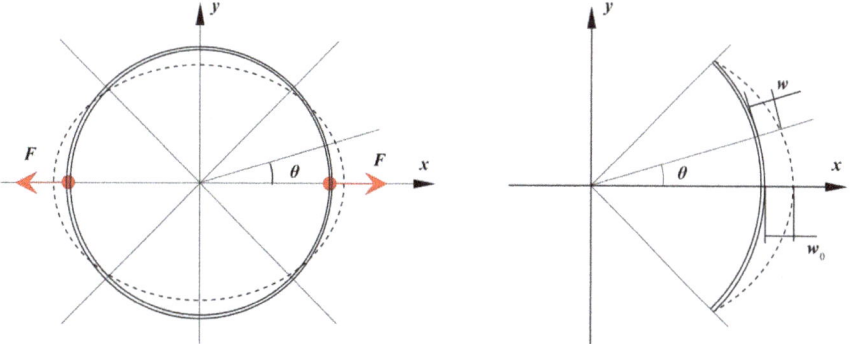

**Fig. 3.7** Distribution of force on the ring and its radial displacement

Considering that the ring is an elastomer whose radial and tangential displacement are both small, generated near its equilibrium, and meet geometric constraint conditions, the variational method in elasticity mechanics is used to calculate undetermined coefficients $a_n$ and $b_n$ in Eq. (3.23) based on the principle of virtual displacement. Let the virtual displacement of the ring under driving force $F$ be as follows (only when $\theta = 0$ and $\theta = \pi$):

$$\delta w = Rn \cos n\theta \delta b_n$$
$$= Rn \cos(n0)\delta b_n + Rn \cos(n\pi)\delta b_n \tag{3.26}$$

According to the principle of virtual work, all of the work done by driving force $F$ within its virtual displacement range is converted into the strain energy of the ring under quasi-static conditions, therefore:

$$2\left(\frac{1}{2}F \cdot \delta w_n\right) = \delta V \tag{3.27}$$

According to Eq. (3.24), the variation of the ring's strain energy $V$ can be evaluated as follows:

$$\delta V = \frac{\partial V}{\partial b_n}\delta b_n = \frac{\pi E I_z}{R}b_n\left(n^3 - n\right)^2 \delta b_n \tag{3.28}$$

Equation (3.28) is the substituted into Eq. (3.27), deriving the following statement:

$$\frac{\pi E I_z}{R}b_n\left(n^3 - n\right)^2 \delta b_n = FRn(1 + \cos n\pi)\delta b_n \tag{3.29}$$

Undetermined coefficient $b_n$ can be evaluated as follows:

$$b_n = \frac{FR^2}{\pi E I_z}\frac{(1 + \cos n\pi)}{n\left(n^2 - 1\right)^2} \tag{3.30}$$

The parity of $n$ is discussed, with the following statement being derived:

$$\begin{cases} b_n = 0, \, n \text{ is even} \\ b_n = \frac{FR^2}{\pi E I_z}\frac{2}{n\left(n^2-1\right)^2}, \, n \text{ is odd} \end{cases} \tag{3.31}$$

Equation (3.31) is substituted into Eq. (3.25), deriving the following expression for the displacements of the ring generated by the driving force:

$$\begin{cases} v = \frac{FR^3}{\pi EI_z} \sum_{n=2m}^{\infty} \frac{2}{n(n^2-1)^2} \sin n\theta \\ \quad = \frac{FR^3}{\pi EI_z} \left( \frac{1}{9} \sin 2\theta + \frac{1}{450} \sin 4\theta + \frac{1}{3675} \sin 6\theta + \cdots \right) \\ w = \frac{FR^3}{\pi EI_z} \sum_{n=2m}^{\infty} \frac{2}{(n^2-1)^2} \cos n\theta \\ \quad = \frac{FR^3}{\pi EI_z} \left( \frac{2}{9} \cos 2\theta + \frac{2}{225} \cos 4\theta + \frac{2}{1225} \cos 6\theta + \cdots \right) \end{cases} \tag{3.32}$$

All terms in the above equation after $n = 4$ can be ignored. An approximate displacement expression can be derived as follows:

$$\begin{cases} v = \frac{FR^3}{9\pi EI_z} \sin 2\theta \\ w = \frac{2FR^3}{9\pi EI_z} \cos 2\theta \end{cases} \tag{3.33}$$

According to Eq. (3.33), the radial displacement of the ring where $\theta = 0$ and $\theta = \pi$ can be determined as follows:

$$w_0 = w_\pi = \frac{2FR^3}{9\pi EI_z} \tag{3.34}$$

Radial tensile stiffness of the ring can then be evaluated as follows:

$$k_w = \frac{F}{w_0} = \frac{9\pi EI_z}{2R^3} \tag{3.35}$$

Equation (3.31) is substituted into Eq. (3.24), deriving an expression for the strain energy of the ring under the driving force:

$$V = \frac{\pi}{2} \frac{EI_z}{R} \sum_{n=2}^{\infty} (n^3 - n)^2 \left( \frac{FR^2}{\pi EI} \frac{2}{n(n^2-1)^2} \right)^2 = \frac{2F^2R^3}{\pi EI_z(n^2-1)^2} \tag{3.36}$$

With all terms after $n = 4$ being ignored, the following approximate strain energy expression is derived:

$$V = \frac{2F^2R^3}{9\pi EI_z} \tag{3.37}$$

From the above analysis, driven by the driving force where $\theta = 0$ and $\theta = \pi$, the radial displacement distribution of the ring is approximately elliptical, as described in Eq. (3.33).

## 3.2.2  Analysis of Axial Stiffness

In the axial vibrations of the ring, the vibration direction is consistent with the height axis of the ring. The ring can ten be treated as being equivalent to a thin-walled beam. The height of the beam is the height of the ring, and the length of the beam is the circumference of the ring.

To easily demonstrate the relationship between the axial displacement of the ring in the CVG and the driving force, a semicircular curved beam model can be built for the ring according to its structural characteristics, as shown in Fig. 3.8a.

Considering that the axial displacement of the ring is circumferentially symmetrical, half of the length of the ring is taken to analyze the force borne by it, as shown in Fig. 3.8b. It's assumed that Point $A$ and Point $B$ on the ring, i.e., locations 0 and $\pi$, are restrained ends. At this time, the ring can be treated as being equivalent to a curved beam with double-clamped ends.

The curved beam is subjected to external force $F$ between both ends $A$ and $B$, so the beam is subjected to 6 support reaction forces, which are $F_A$, $F_B$, $N_A$, $N_B$, $M$, and $M_B$ respectively. Under the condition of minor deformation, the centroid of the cross-section is infinitesimally displaced along the circumferential direction of the ring, so the horizontal support reaction forces $F_A$ and $F_B$ are also very small and can be ignored. The following force and moment equilibrium equation is established for the ring:

$$\begin{cases} N_A + N_B = F \\ M_B + M_B = FR \end{cases} \tag{3.38}$$

From the symmetry of the force borne by the ring:

$$\begin{cases} N_A = N_B = F/2 \\ M_B = M_B = FR/2 \end{cases} \tag{3.39}$$

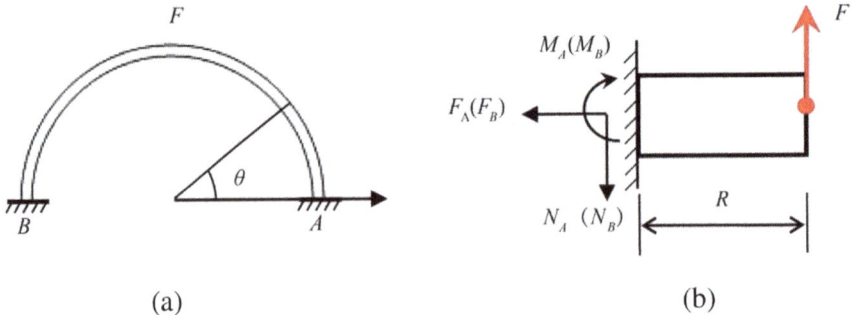

(a)                                             (b)

**Fig. 3.8** Diagram of symmetrical axial force on the ring **a** vertical view; **b** right-side view

According to the support reaction force and support moment on both ends of the above beam, the shear force $Q$ can be evaluated at each point on the ring:

$$Q(\theta) = \begin{cases} -F/2, 0 < \theta < \frac{\pi}{2} \\ F/2, \frac{\pi}{2} < \theta < \pi \end{cases} \tag{3.40}$$

According to the moment balance equation of each ring section, the flexural moment can be evaluated at each point on the ring:

$$M(\theta) = \begin{cases} M_A \cos\theta - \frac{FR}{2}\sin\theta, 0 < \theta < \frac{\pi}{2} \\ M_B \cos(\pi - \theta) - \frac{FR}{2}\sin(\pi - \theta), \frac{\pi}{2} < \theta < \pi \end{cases} \tag{3.41}$$

Equation (3.39) can then be substituted into Eq. (3.41), revealing the distribution of the flexural moment at each point on the ring:

$$M(\theta) = \begin{cases} \frac{FR}{2}(\cos\theta - \sin\theta), 0 < \theta < \frac{\pi}{2} \\ \frac{FR}{2}(-\cos\theta - \sin\theta), \frac{\pi}{2} < \theta < \pi \end{cases} \tag{3.42}$$

After the regularity of distribution of the flexural moment at each point on the ring is revealed, the curved beam is flattened into a straight beam to perform more intuitive analysis. The distribution of the shear force and flexural moment is as follows at each point on the straight beam:

$$Q(x) = \begin{cases} -\frac{F}{2}, 0 < x < \frac{\pi R}{2} \\ \frac{F}{2}, \frac{\pi R}{2} < x < \pi R \end{cases}, \quad M(x) = \begin{cases} \frac{FR}{2}\left(\cos\frac{x}{R} - \sin\frac{x}{R}\right), 0 < x < \frac{\pi R}{2} \\ \frac{FR}{2}\left(-\cos\frac{x}{R} - \sin\frac{x}{R}\right), \frac{\pi R}{2} < x < \pi R \end{cases} \tag{3.43}$$

After the curved beam is flattened along the ring, the shear force and flexural moment distributed along the length axis of the straight beam is diagrammed in Fig. 3.9.

The rotation equation and deflection equation of the beam are derived by the integration method:

$$\begin{cases} \varphi(x) = \int \frac{M(x)}{EI_r}dx + C \\ u(x) = \iint \left(\frac{M(x)}{EI_r}dx\right)dx + Cx + D \end{cases} \tag{3.44}$$

where $C$ and $D$ are integral constants determined by the boundary conditions of the beam, and $I_r$ is the moment of inertia of the ring section around the centerline of the ring parallel to its $r$ axis.

Equation (3.43) is substituted into Eq. (3.44) to evaluate the rotation angle and deflection of the beam in the range of zero to $\pi/2$, i.e., and in the range of zero to $\pi R/2$ after flattening, with the following statement being derived:

**Fig. 3.9** Distribution
diagram of the shear force
and flexural moment on the
ring

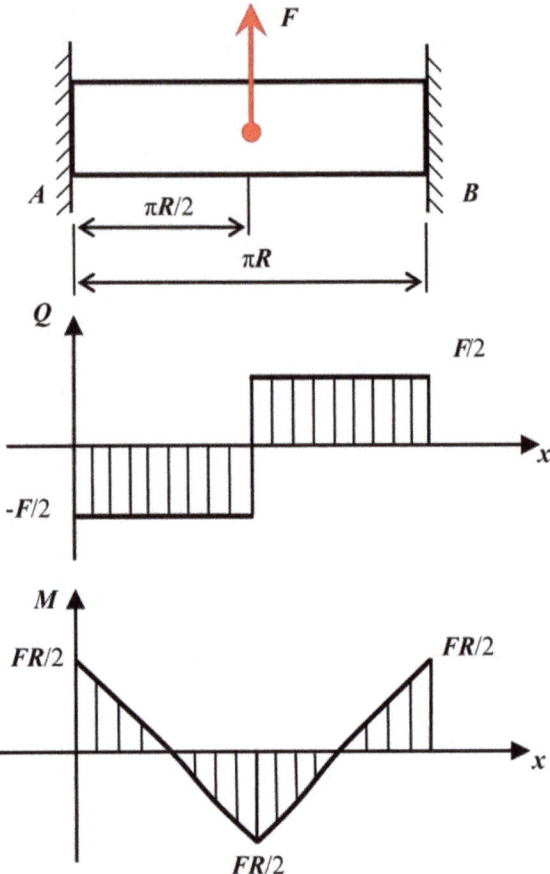

$$\begin{cases} \varphi(x) = \frac{FR^2}{2EI_r}\left(\sin\frac{x}{R} + \cos\frac{x}{R}\right) + C \\ u(x) = \frac{FR^3}{2EI_r}\left(-\cos\frac{x}{R} + \sin\frac{x}{R}\right) + Cx + D \end{cases} \tag{3.45}$$

The boundary conditions of the beam, in line with its rotation angle and deflection at Point $A$, are as follows:

$$\varphi(0) = 0, u(0) = 0 \tag{3.46}$$

Equation (3.46) is substituted into Eq. (3.45), resulting in the following solution:

$$C = -\frac{FR^2}{2EI_r}, D = \frac{FR^3}{2EI_r} \tag{3.47}$$

Therefore, the rotation angle and deflection equations of the beam in the range of 0 to $\pi R/2$ are as follows:

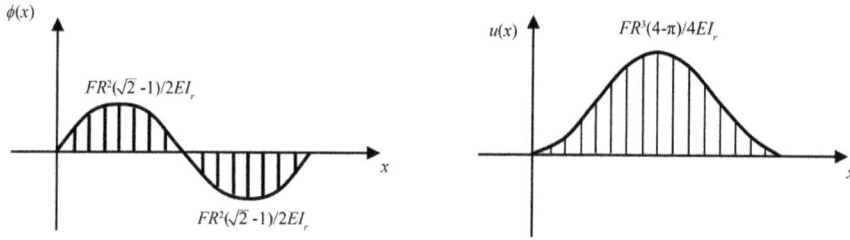

**Fig. 3.10**  Diagram of the ring's rotation and deflection curves

$$\begin{cases} \varphi(x) = \frac{FR^2}{2EI_r}\left(\sin\frac{x}{R} + \cos\frac{x}{R}\right) - \frac{FR^2}{2EI_r} \\ u(x) = \frac{FR^3}{2EI_r}\left(-\cos\frac{x}{R} + \sin\frac{x}{R}\right) - \frac{FR^2}{2EI_r}x + \frac{FR^2}{2EI_r} \end{cases} \tag{3.48}$$

Similarly, the rotation and deflection equations of the beam in the range of $\pi R/2$ to $\pi R$ are as follows:

$$\begin{cases} \varphi(x) = \frac{FR^2}{2EI_r}\left(-\sin\frac{x}{R} + \cos\frac{x}{R}\right) + \frac{FR^2}{2EI_r} \\ u(x) = \frac{FR^3}{2EI_r}\left(\cos\frac{x}{R} + \sin\frac{x}{R}\right) + \frac{FR^2}{2EI_r}x + \frac{FR^3}{2EI_r} - \frac{F\pi R^3}{2EI_r} \end{cases} \tag{3.49}$$

A diagram of rotation and deflection curves distributed along the length of the beam is plotted and shown in Fig. 3.10.

Based on Eq. (3.49), the deflections at points 0, $\pi R/4$, $\pi R/2$, and $3\pi R/4$ on the beam are:

$$\begin{cases} u(0) = 0, \, u(\pi R) = 0, \, u\left(\frac{\pi R}{4}\right) = \frac{FR^3}{2EI_r} - \frac{F\pi R^3}{8EI_r} \\ u\left(\frac{\pi R}{2}\right) = \frac{FR^3}{EI_r} - \frac{F\pi R^3}{4EI_r}, \, u\left(\frac{3\pi R}{4}\right) = \frac{FR^3}{2EI_r} - \frac{F\pi R^3}{8EI_r} \end{cases} \tag{3.50}$$

Since there is an external force acting at 0, $\pi R/2$, and $\pi R$, considering the symmetry of the stress and deformation of the ring, the deflection of the beam should be equal to 0 at $\pi R/4$ and $3\pi R/4$. Therefore, to intuitively reflect the state of deformation of the beam, the deflection curve function can be modified as follows:

$$\begin{cases} u(x) = \frac{FR^3}{2EI_r}\left(-\cos\frac{x}{R} + \sin\frac{x}{R}\right) - \frac{FR^2}{2EI_r}x + \frac{F\pi R^3}{8EI_r}, 0 < x < \frac{\pi R}{2} \\ u(x) = \frac{FR^3}{2EI_r}\left(\cos\frac{x}{R} + \sin\frac{x}{R}\right) + \frac{FR^2}{2EI_r}x - \frac{3F\pi R^3}{8EI_r}, \frac{\pi R}{2} < x < \pi R \end{cases} \tag{3.51}$$

The axial deflection curve function expression (3.51) of the ring is expressed in polar coordinates, with the following statement able to be derived:

$$\begin{cases} u(\theta) = \frac{FR^3}{2EI_r}(-\cos\theta + \sin\theta) - \frac{FR^3}{2EI_r}\theta + \frac{F\pi R^3}{8EI_r}, 0 < \theta < \frac{\pi}{2} \\ u(\theta) = \frac{FR^3}{2EI_r}(\cos\theta + \sin\theta) + \frac{FR^3}{2EI_r}\theta - \frac{3F\pi R^3}{8EI_r}, \frac{\pi}{2} < \theta < \pi \end{cases} \tag{3.52}$$

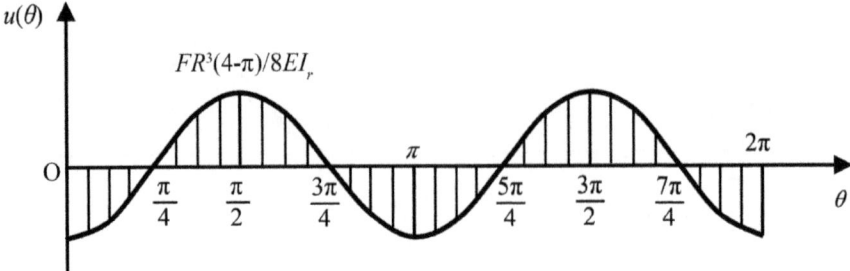

**Fig. 3.11** Diagram of the axial deflection distribution of the ring

Considering the symmetrical distribution of the structure, the stress, axial deflection, and deflection distribution of the ring in the full circumference range can be diagramed and shown in Fig. 3.11.

As can be seen in Fig. 3.11, the deflection distribution of the ring in the full circumference range resembles a cosine function with a period of $\pi$. The deflection function in the range of 0-$\pi$/2 Eq. (3.52) can be rewritten as follows:

$$
\begin{aligned}
u(\theta) &= \frac{FR^3}{2EI_r}(-\cos\theta + \sin\theta) - \frac{FR^3}{2EI_r}\theta + \frac{F\pi R^3}{8EI_r} \\
&= \frac{FR^3(4-\pi)}{8EI_r}\left[\frac{4}{(4-\pi)}(-\cos\theta + \sin\theta - \theta) + \frac{\pi}{(4-\pi)}\right]
\end{aligned}
\tag{3.53}
$$

A comparison can be made between the normalized deflection function in the range of 0-$\pi$/2, i.e.,

$$
u_1(\theta) = \frac{4}{(4-\pi)}(-\cos\theta + \sin\theta - \theta + 1)
\tag{3.54}
$$

and function cos2$\theta$, as shown in Fig. 3.12.

As shown by Fig. 3.12, there is a very small difference in the numerical value between $u_1(\theta)$ and cos2$\theta$ in the range of 0-$\pi$/2, and the maximum absolute error appears near $\theta=\pi$/8 and $\theta=3\pi$/8, and is equal to 0.01547. The maximum relative error appears near $\theta=\pi$/4, and is equal to 3.61%. Therefore, the axial deflection function in the full circumference range of the ring is simplified to the following cosine function:

$$
u(\theta) = -\frac{FR^3(4-\pi)}{8EI_r}\cos 2\theta
\tag{3.55}
$$

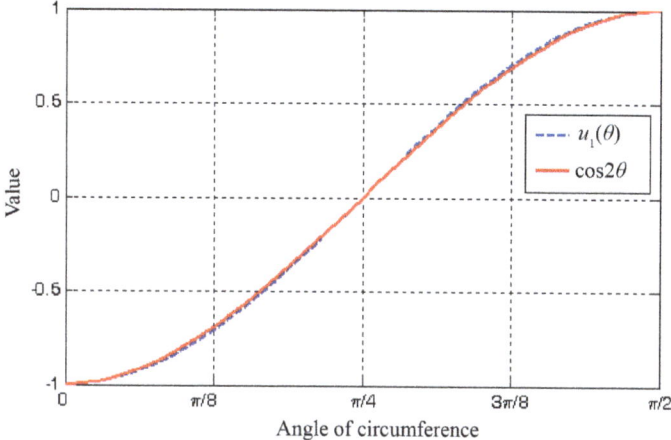

**Fig. 3.12** A comparison of the normalized function and cosine function of the ring's axial deflection

According to Eq. (3.55), the axial displacement of the ring at $\theta = 0$ and $\theta = \pi$ is:

$$u_0 = u_\pi = \frac{FR^3(4 - \pi)}{8EI_r} \tag{3.56}$$

The axial stiffness of the ring can therefore be expressed as follows:

$$k_u = \frac{F}{u_0} = \frac{8EI_r}{(4 - \pi)R^3} \tag{3.57}$$

## 3.3 Mechanical Modeling of Resonators

The mechanical modeling of CVG resonators is the foundation for studying the resonators' dynamic characteristics. Research on the mechanical characteristics of resonator shell structures is mostly based on classic plate and shell vibration theory. In the present section, we discuss the stress and strain of the resonators in their working modes, allowing for the creation of a mechanical parameter model for both the concentration stiffness and lumped mass of the resonator.

### 3.3.1 Concentration Stiffness Model

For problems involving the stiffness of CVG resonators, a type of thinned-wall elastic component, the relationship between the stiffness of each key component of the

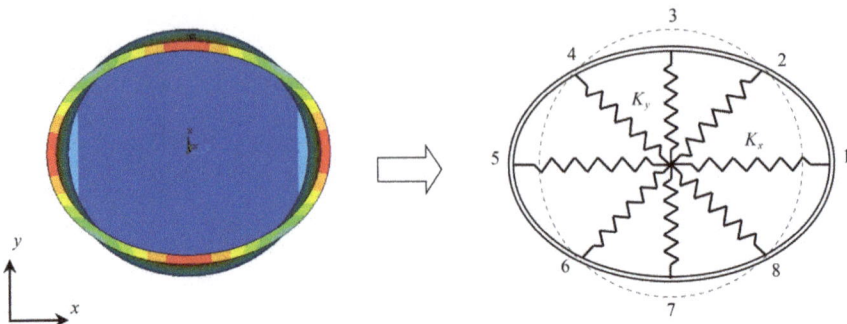

**Fig. 3.13**  Concentration stiffness model of resonators in the *xoy* plane

resonator and its static deformation can be studied in detail from a static perspective, ignoring the impact of self-weight on static deformation. To facilitate the analysis of the relationship between the driving force of the resonator and its displacement in all directions, a concentration stiffness model is built for the ring and is shown in Fig. 3.13. This is structured according to the piezoelectric driving principle of the CVG and research conclusions on the radial and axial stiffness of the ring, as well as the axial symmetry of the resonator's stress and deformation.

Figure 3.13 shows the concentration stiffness model of the resonator in the *xoy* plane. This model primarily reflects the positional relationship between the radial displacement distribution of the shell wall of the resonator and its driving moment. According to the analysis of the radial stiffness of the ring, eight virtual springs can be set at the antinodal and nodal points of the ring's radial displacement. In the concentration stiffness model of the resonator shown in Fig. 3.13, when the resonator is subjected to symmetric forces from Point 1 and Point 5, or from Point 3 and Point 7, springs 1, 3, 5, and 7 all contribute. When the resonator is subjected to symmetrical forces from Point 2 and Point 4 or from Point 6 and Point 8, springs 2, 4, 6, and 8 all contribute.

A concentration stiffness model is built in the *xoz* plane [i.e., section $\theta = 0$ in the cylinder coordinate system $(r, \theta, z)$] where the piezoelectric electrodes are located, and are shown in Fig. 3.14.

In the concentration stiffness model of the resonator shown in Fig. 3.14, the two springs at Point A reflect the relationship between the displacement of the resonator's shell wall and its acting force, while the torsional spring at Point B reflects the displacement of the bottom of the resonator and the applied moment. In this model, *x*-axial tensile stiffness $K_x$ is primarily determined by the radial stiffness of the shell wall of the resonator. The *z*-axial tensile stiffness $K_z$ is primarily determined by the axial stiffness of the resonator's shell wall; the *y*-axial flexural stiffness $K_\varphi$ is primarily determined by the flexural stiffness of the bottom of the resonator.

In the same way, a concentration stiffness model can be built in the longitudinal section *yoz* [i.e., the section of $\theta = \pi/2$ in the cylindrical coordinate system $(r, \theta, z)$] in the same state of static balance, as shown in Fig. 3.15.

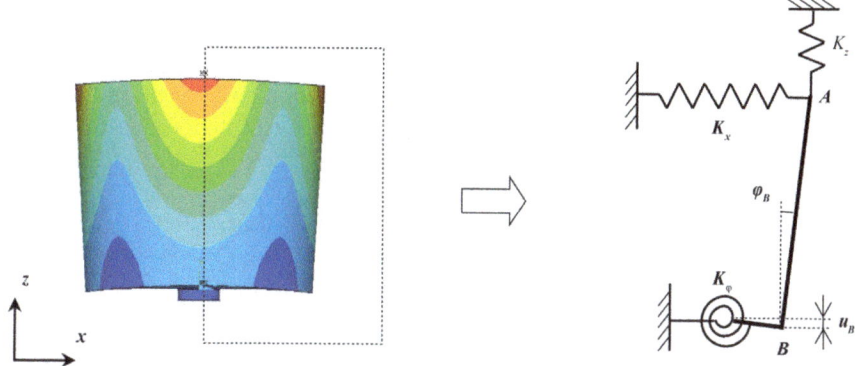

**Fig. 3.14**  Concentration stiffness model of resonators in the $xoz$ plane

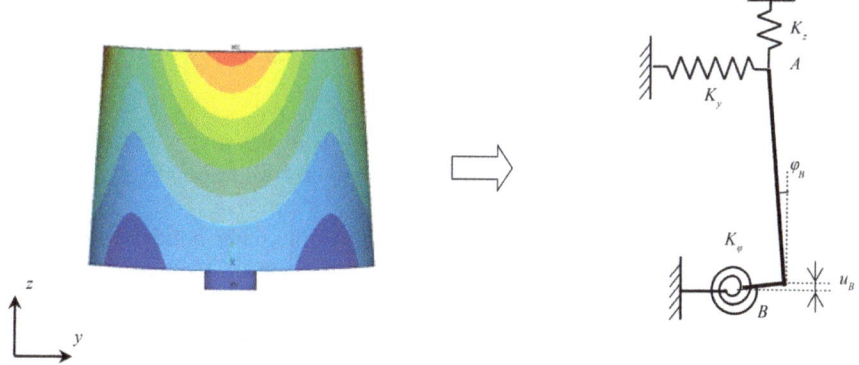

**Fig. 3.15**  Concentration stiffness model of the resonator in the $yoz$ plane

According to the results of the resonator model analysis, when the shell wall of the resonator vibrates, the deflection of its generatrix is negligible, so the generatrix of the resonator's shell can be considered highly rigid. At intersection $B$ between the outer bottom edge of the resonator and its bottom end, the rotation angle at Point $B$ on the corresponding bottom deflection curve and the rotation angle of the generatrix of the shell wall are both $\varphi_B$.

The flexural moment of reaction on the outer bottom edge of the resonator from the shell wall of the resonator is mostly generated by the radial displacement of the resonator's shell wall. To evaluate the flexural moment of reaction on Point $B$ on the outer bottom edge of the resonator from the shell wall of the resonator, the shell wall of the resonator can be discretized into a number of rings along its height axis. As can be seen from the radial tensile stiffness formula of the rings, the radial stiffness of the resonator's shell wall along its micro-height segments is equal to:

$$dk_w = \frac{9\pi E \ t^3 dz}{2R^3} \frac{1}{12} = \frac{3\pi E t^3}{8R^3} dz \tag{3.58}$$

Therefore, the clockwise flexural moment of reaction on Point $B$ on the outer bottom edge of the resonator from the shell wall of the resonator is equal to:

$$
\begin{aligned}
M_B &= \int_0^H z \cdot z\varphi_B dk_w = \int_0^H \frac{3\pi E t^3}{8R^3} z^2 \varphi_B dz \\
&= \frac{\pi E t^3}{8R^3} H^3 \varphi_B \\
&= K_x H\varphi_B \cdot H
\end{aligned}
\tag{3.59}
$$

where $K_x$ represents the equivalent radial tensile stiffness of the shell wall of the resonator, expressed as:

$$K_x = \frac{\pi E t^3 H}{8R^3} \tag{3.60}$$

Because the generatrix of the resonator's shell wall has a very small angle of rotation, the axial displacement of the resonator's shell wall caused by the tilting of the generatrix can usually be ignored.

The reaction force of the resonator shell wall to the outer bottom edge of the resonator is primarily generated by the axial displacement of the resonator's shell wall. To evaluate the reaction on the bottom of the resonator to the resonator's shell wall, the latter can be treated as being equivalent to a full-circle curved beam. Based on the axial stiffness formula of the rings, when each of the points on the generatrix of the resonator's shell wall is subjected to rigid displacement $u$ with the outer bottom edge of the resonator, the resonator's shell generates a $z$-axial reaction on Point $B$ on the outer bottom edge of the resonator:

$$
\begin{aligned}
F_B &= k_u \cdot u_B = \frac{8E I_r u_B}{(4 - \pi)R^3} \\
&= K_z \cdot u_B
\end{aligned}
\tag{3.61}
$$

where $K_z$ represents the equivalent axial tensile stiffness of the shell, which can be approximately expressed as:

$$K_z = \frac{8E I_r}{(4 - \pi)R^3} = \frac{8E t H^3}{3(4 - \pi)R^3} \tag{3.62}$$

The bottom of the resonator is approximately equivalent to eight flexural piezo-electric beams, and Point $B$ on the terminal end of the piezoelectric beams is subjected to a moment. Based on the beam flexure theory of material mechanics, the relationship between the terminal rotation angle of the cantilever beam and the driving moment can be revealed as follows:

$$\begin{cases} \varphi_B = \frac{M_B(R-R_0)^2}{2EI_\theta} \\ u_B = \frac{M_B(R-R_0)}{EI_\theta} \end{cases} \tag{3.63}$$

where $I_\theta$ represents the polar moment of inertia of the equivalent beam at the bottom.

It can be thus seen that the curvature of the equivalent flexural piezoelectric beam at the bottom of the resonator can be equivalent to the rotation of Point $B$ around the center of torsional spring $K_\varphi$. Its rotation radius is as follows:

$$R_\varphi = \frac{u_B}{\varphi_B} = \frac{R - R_0}{2} \tag{3.64}$$

According to Eq. (3.63), the equivalent flexural stiffness of the flexural piezoelectric beam at the bottom of the resonator can be calculated as follows:

$$K_\varphi = \frac{EI_\theta}{R - R_0} \tag{3.65}$$

### 3.3.2 Lumped Mass Model

For a vibratory element like a resonator, the impact of mass distribution at all key parts on its effective inertial mass usually need to be analyzed, as this is the key to studying the Coriolis effect within the resonator. To facilitate the analysis of the relationship between mass distribution and the inertial mass, a lumped mass model is built for the resonator vibrating on spring $K_x$ (in the $xoy$ plane). By setting the maximum radial displacement on the top of the resonator's shell wall as a generalized coordinate based on the preceding research on the natural vibrations of the resonator in its working modes as well as its structural characteristics. This method is shown in Fig. 3.16.

Judging from the analysis of the natural vibrations of the resonator, the radial and tangential vibrational energy of the resonator's shell wall can be expressed as follows:

$$T_{cx} = \int_0^{2\pi} \int_0^H \frac{1}{2}\left(\dot{w}^2(\theta, z) + \dot{v}^2(\theta, z)\right)\rho t_2 dz R d\theta \tag{3.66}$$

The vibration velocity (see Sect. 4.1) expression of the shell wall of the resonator is substituted into Eq. (3.66), with the following equation able to be derived:

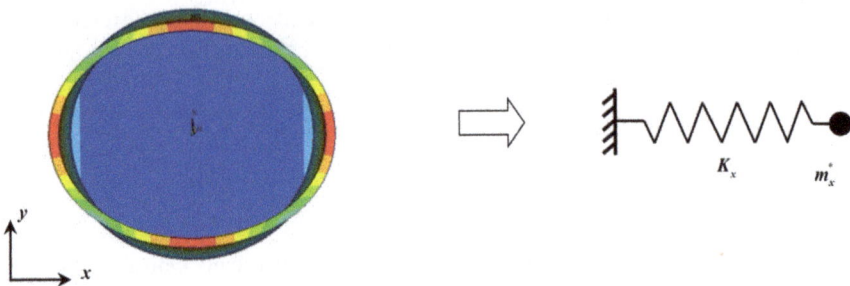

**Fig. 3.16** Lumped mass model of the resonator in the *xoy* plane

$$T_{cx} = \int_0^{2\pi} \int_0^H \frac{1}{2}\left(\frac{z}{H}\right)^2 \left((w_0\omega \cos \omega t \cos 2\theta)^2 + (v_0\omega \cos \omega t \sin 2\theta)^2\right)\rho t_2 dz R d\theta$$

$$= \frac{1}{2}(w_0\omega \cos \omega t)^2 \frac{5\pi\rho RHt}{12}$$

$$= \frac{1}{2}(w_0\omega \cos \omega t)^2 m_{cx}^* \tag{3.67}$$

where $m_{cx}^*$ represents the effective inertial mass of the shell on spring $K_x$, being expressed as:

$$m_{cx}^* = \frac{5\pi\rho RHt}{12} \tag{3.68}$$

According to the equation for the vibrations at the bottom of the resonator, the vibration velocity in the *xoy* plane can be ignored, so,

$$T_{bx} = 0, \quad m_{bx}^* = 0 \tag{3.69}$$

The effective inertial mass of the resonator on spring $K_x$ can be calculated according to Eqs. (3.68) and (3.69). The results are as follows:

$$m_x^* = m_{cx}^* + m_{bx}^* = \frac{5\pi\rho RHt}{12} \tag{3.70}$$

Similarly, by setting the maximum axial displacement on the top of the shell wall of the resonator as a generalized coordinate, a lumped mass model is established for vibrations on spring $K_z$ (in the *xoz* plane), as shown in Fig. 3.17.

From the analysis on the natural vibrations of the resonator, the axial vibrational energy of the shell wall of the resonator can be expressed as follows:

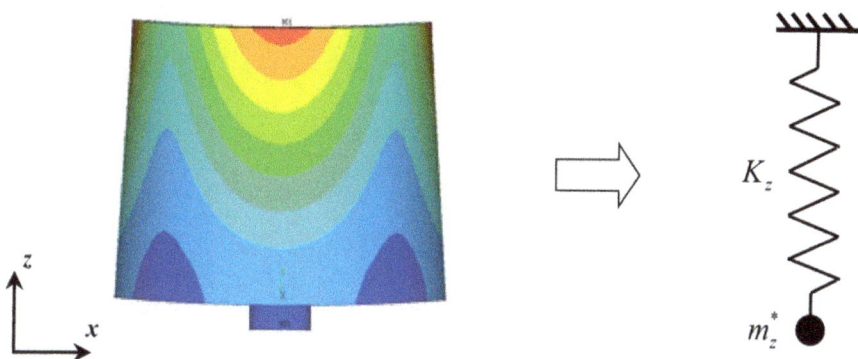

**Fig. 3.17** Lumped mass model of the resonator in the *xoz* plane

$$T_{cz} = \int_0^{2\pi} \int_0^H \frac{1}{2}\dot{u}^2(\theta, z)\rho t_2 dz R d\theta \tag{3.71}$$

The vibration velocity expression (4.7) of the shell wall of the resonator is substituted into Eq. (3.71), with the following able to be derived:

$$T_{cz} = \int_0^{2\pi} \int_0^H \frac{1}{2}(u_0\omega \cos \omega t \cos 2\theta)^2 \rho t_2 dz R d\theta$$
$$= \frac{1}{2}(u_0\omega \cos \omega t)^2 \pi R\rho H t = \frac{1}{2}(w_0\omega \cos \omega t)^2 m_{cz}^* \tag{3.72}$$

where $m_{cz}^*$ represents the effective inertial mass of the shell wall of the resonator vibrating on spring $K_z$, expressed as:

$$m_{cz}^* = \pi R\rho H t \tag{3.73}$$

The axial vibrational energy at the bottom of the resonator can be expressed as follows:

$$T_{cz} = \int_0^{2\pi} \int_{R_0}^R \frac{1}{2}\dot{u}^2(\theta, r)\rho t_3 r dr d\theta \tag{3.74}$$

The vibration velocity expression of the bottom of the resonator is substituted into Eq. (3.70), with the following equation able to be derived:

$$T_{bz} = \int_0^{2\pi} \int_{R_0}^{R} \frac{1}{2} \left( \left( \frac{r - R_0}{R - R_0} \right)^2 u_0 \omega \cos \omega t \cos 2\theta \right)^2 \rho t_3 r dr d\theta$$

$$= \frac{1}{2} (u_0 \omega \cos \omega t)^2 \pi \rho t_3 \frac{5R^2 - 4RR_0 - R_0^2}{30}$$

$$= \frac{1}{2} (w_0 \omega \cos \omega t)^2 m_{bz}^* \tag{3.75}$$

where $m_{bz}^*$ represents the effective inertial mass of the bottom of the resonator vibrating on spring $K_z$, expressed as:

$$m_{bz}^* = \pi \rho t_3 \frac{5R^2 - 4RR_0 - R_0^2}{30} \tag{3.76}$$

The effective inertial mass of the resonator vibrating on spring $K_z$ can be approximated as:

$$m_z^* = m_{cz}^* + m_{bz}^* = \pi \rho R H t \tag{3.77}$$

Based on the above, a lumped mass model is built for the resonator, as shown in Fig. 3.18.

The vibration characteristics of the resonator show that the deflection angle $\varphi B$ of the shell wall and the bottom of the resonator is extremely small. So, in Fig. 3.18, there is both $x$-axial vibration velocity and $z$-axial vibration velocity at lumped mass point $A$, while there is only $z$-axial vibration velocity at lumped mass point $B$. Therefore,

$$\begin{cases} m_A = m_x^* = \frac{5\pi \rho R H t}{12} \\ m_B = m_z^* - m_x^* = \frac{7\pi \rho R H t}{12} \end{cases} \tag{3.78}$$

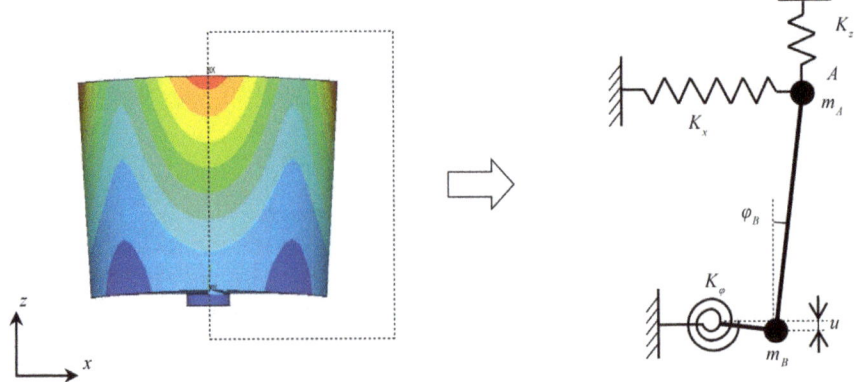

**Fig. 3.18** Lumped mass model of the resonator

# References

1. Warburton, G. (1965). Vibration of thin cylindrical shells. *Journal of Mechanical Engineering Science, 7*(4), 399–407.
2. Salahifar, R., & Mohareb, M. (2010). Analysis of circular cylindrical shells under harmonic forces. *Thin-Walled Structures, 48*(7), 528–539.
3. Wu, L. (1989). *Theory of plates and shells.* Shanghai: Shanghai Jiao Tong University Press.
4. Loveday, P. W., & Rogers, C. A. (1998). Free vibration of elastically supported thin cylinders including gyroscopic effects. *Journal of Sound and Vibration, 217*(3), 547–562.
5. Tao, Y. (2011). *Research on the key technologies of cup-shaped wave gyros.* Changsha: National University of Defense Technology.
6. Bickford, W. B., & Reddy, E. D. (1985). On the in-plane vibrations of rotating rings. *Journal of Sound and Vibration, 101*(1), 10.

# Chapter 4
# Dynamic Analysis and Modeling of CVGs

Dynamic modeling and analysis are extremely important to CVG research and of great significance for studying gyroscopic sensitivity, structural optimization, and control method. This chapter presents the kinetic equations established for the resonator's lumped parameter models. After calculating the resonator's dynamic response, the gyroscope's output and sensitivity can be analyzed.

## 4.1 Modal Analysis of Resonators

Eigenmode is an inherent property of both multi-degree-of-freedom (MDOF) linear systems and continuous elastomer systems. It can be expressed in both the eigenvalue (natural frequency) and characteristic vector (principal mode) of a system, and can be used to describe the system's vibration characteristics in terms of both space and time. The working modes of a CVG refer to the drive mode and sense mode of its resonator, which together provide the basis for the generation of gyroscopic effects. The modal characteristics of the resonator largely determine the design of the drive and sensor units. They also greatly influence the sensitivity of the gyroscopes and serve as an important basis for the design of its sensor and control circuits.

A resonator is a continuous elastomer structure, and as such has infinite-order modes in an infinitely wide frequency range, so it can be considered an infinite-degrees-of-freedom (IDOF) system. In CVGs, gyroscopic effects are achieved through the use of two working modes at the same frequency. The working modes of a given gyroscope are in fact equivalent to the forced vibrations of its resonator structure, and its dynamic characteristics largely determine the gyroscope's performance. As is shown in Fig. 4.1, the first eight orders of a resonator's modes show a certain degree of representativeness, and its mode characteristics are listed in Table 4.1.

Normally, the mode superposition method is used to analyze the dynamic response of an MDOF system. In other words, any vibration displacement component $u$ of a

© National Defense Industry Press 2021

X. Wu et al., *Cylindrical Vibratory Gyroscope*, Springer Tracts in Mechanical Engineering, https://doi.org/10.1007/978-981-16-2726-2_4

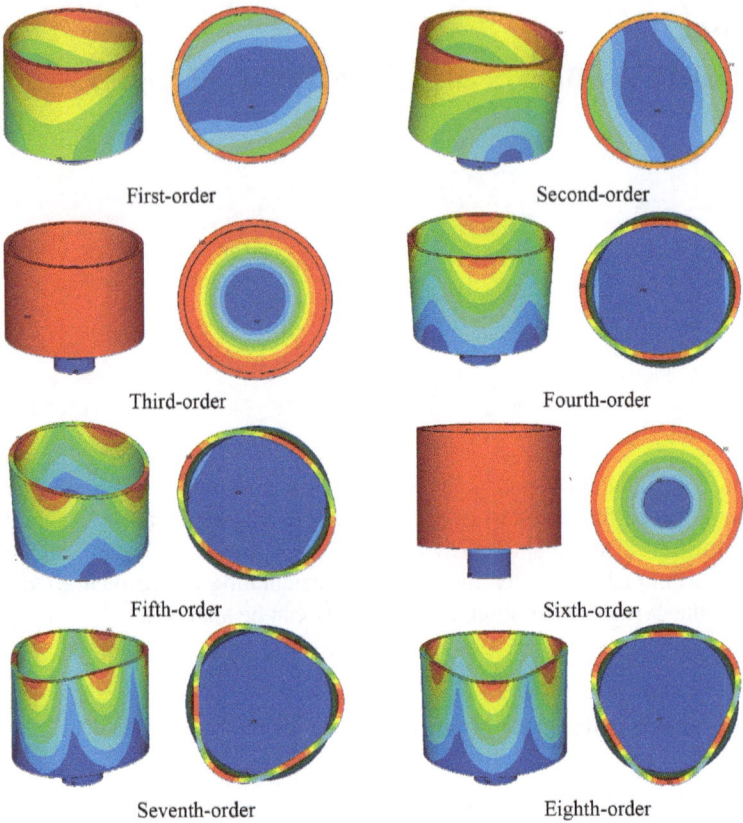

First-order                                          Second-order

Third-order                                          Fourth-order

Fifth-order                                          Sixth-order

Seventh-order                                        Eighth-order

**Fig. 4.1** Contour of various order modes of a CVG resonator

**Table 4.1** Modal frequency and mode characteristics of a CVG resonator

| Modal order | Mode characteristics |
| --- | --- |
| First-order | The shell wall of the resonator has $x$-axial rigid vibrations in the $xoy$ plane |
| Second-order | The shell wall of the resonator has $y$-axial rigid vibrations in the $xoy$ plane |
| Third-order | The shell wall of the resonator has $z$-axial rigid vibrations |
| Fourth-order | The shell wall of the resonator has $x$–$y$-axial "circular-elliptical" elastic vibrations in the $xoy$ plane |
| Fifth-order | The shell wall of the resonator has $x'$-$y'$-axial "circular-elliptical" elastic vibrations in the $xoy$ plane |
| Sixth-order | The shell wall of the resonator has $z$-axial torsional elastic vibrations in the cylindrical plane |
| Seventh-order | The shell wall of the resonator has $x$–$y$-axial "circular-triangular circular" elastic vibrations in the $xoy$ plane |
| Eighth-order | The shell wall of the resonator has $x'$-$y'$-axial "circular-triangular-circular" elastic vibrations in the $xoy$ plane |

structure is evaluated by superimposing each mode component $u_i$ over each other. The displacement of any mode component $u_i$ can be evaluated by multiplying vibration distribution vector $\phi_i$ by mode amplitude $Y_i$, as follows:

$$u_i = \phi_i Y_i \tag{4.1}$$

The total displacement of the structure is evaluated by the sum of the mode components, as follows:

$$u = \phi_1 Y_1 + \phi_2 Y_2 + \cdots \phi_N Y_N = \sum_{i=1}^{N} \phi_i Y_i \tag{4.2}$$

This can be expressed in matrix notation as:

$$u = \Phi Y \tag{4.3}$$

where the function of mode matrix $\Phi$ is transforming generalized coordinate $Y$ into geometric coordinate vector $u$. The generalized elements in vector $Y$ are the canonical coordinates of the structure.

Canonical coordinate transforms can be performed to transform the $N$ coupled linear damped motion equations of the MDOF system, i.e.,

$$m\ddot{u}(t) + c\dot{u}(t) + ku(t) = p(t) \tag{4.4}$$

into $N$ uncoupled motion equations, as follows:

$$\ddot{Y}_i(t) + 2\xi_i \omega_i \dot{Y}_i(t) + \omega_i^2 Y_i(t) = \frac{p(t)}{M_i} \tag{4.5}$$

where $M_i = \phi_i^T m \phi_i$, $P_i(t) = \phi_i^T p(t)$. To solve these uncoupled motion equations, the relevant mode shape distribution vector $\phi_i$ and natural frequency $\omega_i$ need to be evaluated first. The modal damping ratio is usually difficult to determine, but it can be assumed or measured experimentally. After solving $N$ standard single-DOF motion equations, the dynamic responses of the relevant modes can be superimposed to reveal the overall dynamic response of the MDOF system.

For the forced vibration of the resonator generated by the piezoelectric driving force, the mode superposition method indicates that its dynamic response can be expressed as:

$$\begin{bmatrix} u \\ v \\ w \end{bmatrix} = \begin{bmatrix} \phi_{u\_1} Y_{u\_1} \\ \phi_{v\_1} Y_{v\_1} \\ \phi_{w\_1} Y_{w\_1} \end{bmatrix} + \begin{bmatrix} \phi_{u\_2} Y_{u\_2} \\ \phi_{v\_2} Y_{v\_2} \\ \phi_{w\_2} Y_{w\_2} \end{bmatrix} + \cdots \begin{bmatrix} \phi_{u\_N} Y_{u\_N} \\ \phi_{v\_N} Y_{v\_N} \\ \phi_{w\_N} Y_{w\_N} \end{bmatrix} + \cdots = \sum_{i=1}^{\infty} \begin{bmatrix} \phi_{u\_i} Y_{u\_i} \\ \phi_{v\_i} Y_{v\_i} \\ \phi_{w\_i} Y_{w\_i} \end{bmatrix} \tag{4.6}$$

**Fig. 4.2** Modal contour of CVG resonator sections in its working modes **a** *xoy* plane; **b** *xoz* plane; **c** *yoz* plane

Equation (4.6) represents the displacement vector of the resonator in geometric coordinates. It is formed by the dynamic displacements of the resonator in the given order modes. For most load types, a structure's low-order modes generally contribute far more to this structure's dynamic response. Order modes that are closest to the excitation frequency contribute the greatest to the structure's dynamic response, while higher-order modes contribute less to its dynamic response. In addition, the higher-order modes of a complex structure are typically difficult to reliably predict. Therefore, when the mode superposition method is used to study the dynamic response of the resonator to the driving moment, its responses in all high-order modes can be omitted from consideration, while close attention should be paid to the structure's dynamic response in the modes smaller than its working modes. This allows focus to be placed on the problem of forced vibrations at the frequencies corresponding to the resonator's working modes.

The working modes of a CVG resonator are its fourth and fifth-order modes, and the focus of the analysis will be on the characteristics of the resonators' working modes. Figure 4.2a shows a modal nephogram of the top plane (*xoy* plane) of the resonator. Figure 4.2b, c show a modal nephogram of the longitudinal planes which include the maximum radial displacements of the resonator (*xoz* plane and *yoz* plane).

According to the distribution characteristics of modal displacements on all sections shown in Fig. 4.2, the vibration amplitude of each node on the top of the shell, the vibration amplitude of each generatrix node in the shell of the resonator, and the vibration amplitude of each node in the inner radius of the resonator bottom are extracted to calculate the displacement distribution at all of the above key nodes, as shown Fig. 4.3.

As can be seen from the circumferential position ($\theta$) of each point on the inner circle of the shell wall top of the resonator, its radial displacement ($w$), and tangential displacement ($v$), while in a working mode, the distribution characteristics of the radial displacement and tangential displacement of the resonator's shell wall can be demonstrated to be as follows: For circumferential wavenumber $n = 2$, the radial displacement of the shell wall of the resonator accords with the regularity of distribution in $\cos 2\theta$, while its tangential displacement accords with the regularity of distribution in $\sin 2\theta$.

The radial displacement ($w$) of each generatrix point on the shell wall of the resonator and data distributed in the axial direction ($z$) are linearized. As shown in

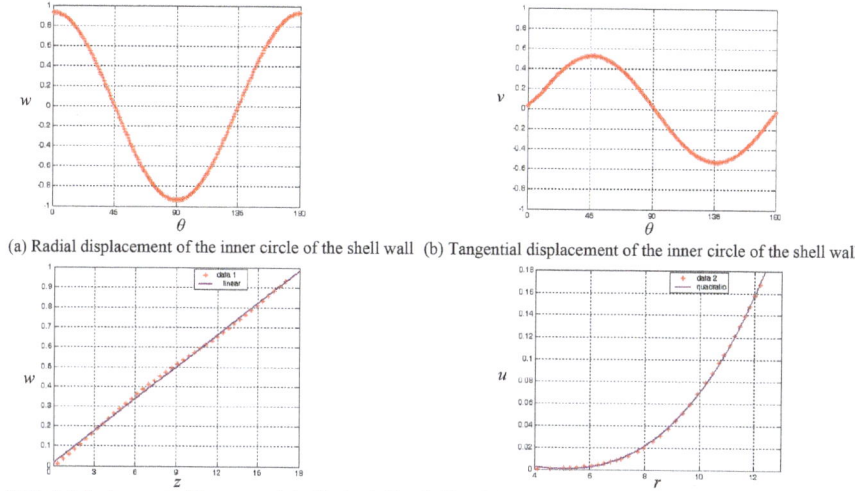

(a) Radial displacement of the inner circle of the shell wall (b) Tangential displacement of the inner circle of the shell wall

(c) Radial displacement of the generatrix inside the shell wall (d) Axial displacement of the inside radius of the bottom

**Fig. 4.3** Displacement distribution on key nodes of the CVG resonator

Fig. 4.3c, the non-linearity is about 1% over a finite length range. Therefore, in a working mode, the generatrix displacement distribution characteristic function $R_m(z)$ can be assumed to be a linear function.

Figure 4.3d shows the results of quadratic curve fitting of the data about the radial position $(r)$ and axial displacement $(u)$ of each point in the inner bottom radius of the resonator. The results show that the data in the figure have smaller variance than the fitted quadratic function, and that the quadratic function is similar to the deflection function of the cantilever beam. Therefore, in a working mode, the displacement distribution function of bottom radius, i.e., the Bessel function, can be simplified into a quadratic deflection function of the cantilever beam. The simplification will make it easier to evaluate the dynamic response of the resonator at its bottom.

In summary, a motion equation for the working modes of a CVG resonator is inductively established based on the results of the finite element modal analysis of the resonator by referencing the classic theory of plates and shells, as follows:

The vibration displacement distribution function of the shell wall in working modes is as follows:

$$\begin{cases} v_c(r, \theta, z, t) = B\frac{z}{H} \sin 2\theta \sin \omega t \\ u_c(r, \theta, z, t) = A \cos 2\theta \sin \omega t \\ w_c(r, \theta, z, t) = C\frac{z}{H} \cos 2\theta \sin \omega t \end{cases} \tag{4.7}$$

The vibration displacement distribution function of the bottom of the resonator in its working modes is as follows:

$$\begin{cases} v_b(r, \theta, z, t) = 0 \\ u_b(r, \theta, z, t) = D\left(\frac{r-R_0}{R-R_0}\right)^2 \cos 2\theta \sin \omega t \\ w_b(r, \theta, z, t) = 0 \end{cases} \tag{4.8}$$

## 4.2  Kinetic Equations of Resonator Drive Mode

According to the results of the lumped concentration stiffness model and lumped mass model of resonators produced in Chap. 3, each point on the generatrix inside the shell wall of the resonator is subjected to rigid displacement and rigid inclination with the outer bottom edge of the resonator under the action of driving moment $M_p$ of the flexural piezoelectric actuator at the bottom of the resonator. The displacement is the deflection $u_B$ of the outer bottom edge of the resonator, while the angle of tilt is the rotation angle $\varphi_B$ of the outer bottom edge of the resonator, so the resonator can be simplified into an equivalent lumped parameter model, as shown in Fig. 4.4 [1].

In the model shown in Fig. 4.4, $M_p$ is the driving moment. The model also contains two lumped mass elements and three concentration stiffness elements. The motion state variables in the model include the radial displacement $w_A$ of Point A on the top of the shell wall of the resonator, and the axial displacement $u_B$ (the deflection of the outer endpoint of the resonator bottom) of Point B on the shell wall bottom of

**Fig. 4.4** Driving moment and equivalent lumped parameter model of a resonator

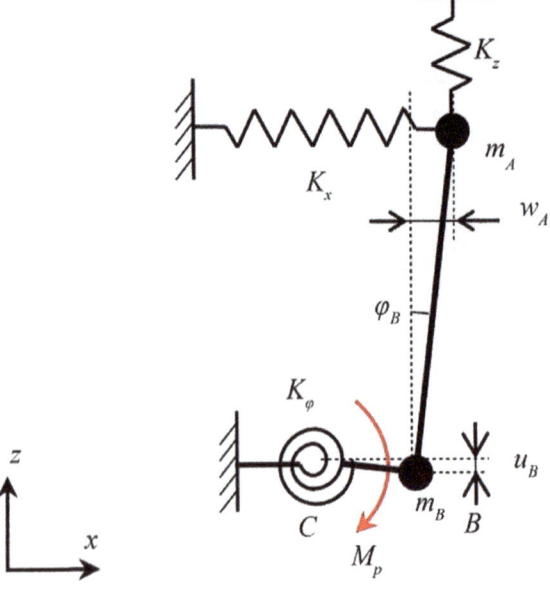

the resonator. Owing to the small angles of deflection and rotation of the resonator bottom, the rotation angle $\varphi_B$ of Point $B$ and the radial displacement $w_A$ of Point $A$ form a relationship, i.e., $w_A = H\varphi_B$. In addition, the relationship between the radial displacement $w_A$ of Point $A$ and the axial displacement $u_B$ of Point $B$ is determined based on Eq. (3.64), as follows:

$$\frac{w_A}{u_B} = -\frac{2H}{R - R_0} \tag{4.9}$$

From Eq. (2.14) in Chap. 2, the driving moment of the flexural piezoelectric actuator can be seen to be not only related to the driving voltage, but also to the flexural deformation state of the flexural piezoelectric actuator. For the purposes of studying the linear vibrations caused by the micro-displacement of the resonator, the impact of the flexural deformation of the flexural piezoelectric actuator on its driving moment can be said to be far less than the impact of the driving voltage on the driving moment. This is because the strain of the piezoelectric electrodes is very small under the restraining action of the shell wall and the bottom of the resonator, and this in turn is caused by the stiffness of the shell wall and the bottom of the resonator being much higher than that of the piezoelectric electrodes. To simplify the analytical process, the second term on the right of Eq. (2.14) can be ignored for the purposes of studying the vibrations caused by the micro-displacement of the resonator. The driving moment of the existing flexural piezoelectric actuator is calculated as follows:

$$M_p(t) = U_p(t)\left(\frac{c_{13}e_{33}}{c_{33}} - e_{31}\right)b_p\frac{h_b + h_p}{2}$$
$$= U_p(t)\Theta_{UM} \tag{4.10}$$

where $\Theta_{UM} = \left(\frac{c_{13}e_{33}}{c_{33}} - e_{31}\right)b_p\frac{h_b+h_p}{2}$, represent the voltage-moment conversion coefficient of the resonator.

As can be seen, the motion of the lumped parameter model of the resonator can be simplified into the rigid motion of two lumped mass elements around the center of the torsional spring at the bottom under the restraining action of three concentration stiffness elements. Thus, in the working modes of a CVG, the lumped parameter model of the resonator can be treated as equivalent to a generalized single-DOF system composed of multiple stiffness elements and multiple mass elements.

According to the direct balancing method offered in the D'Alembert principle, a force equilibrium equation can be established for the resonator based on the inertial force of the lumped mass elements as well as the elastic restraining force and driving force of the concentration stiffness elements in the equivalent lumped parameter model of the resonator. Taking the approximately equivalent torsional spring at the bottom of the resonator as an object of study, in the state shown in Fig. 4.4, the moment of the inertial force and elastic restraining force on the center point $C$ of the torsional spring is expressed as:

$$\begin{cases} M_a = m_A \ddot{w}_A H + (m_A + m_B)\ddot{u}_B \frac{1}{2}(R - R_0) \\ M_k = K_x w_A H + K_z u_B \frac{1}{2}(R - R_0) \end{cases} \tag{4.11}$$

where $M_a$ represents the moment of the inertial force of the lumped mass elements on Point $C$; $M_k$ represents the restoring force of concentration stiffness elements on Point $C$; $\ddot{w}_A$ and $\ddot{u}_B$ are the second derivative of the radial displacement $w_A$ of Point $A$ and the axial displacement $u_B$ of Point $B$ over time, respectively.

Ignoring the impact of resonator damping, a force equilibrium equation is established for the equivalent lumped parameter model of the resonator based on the driving moment of the flexural piezoelectric actuator and the equivalent stiffness formula of the flexural piezoelectric beam, as follows:

$$M_k + M_a + K_\varphi \cdot \varphi_B = M_p \tag{4.12}$$

Equations (4.10) and (4.11) are substituted into Eq. (4.12), deriving a kinetic equation for the equivalent lumped parameter model of the resonator, with the radial displacement $w_A$ of Point $A$ as a generalized coordinate, shown below:

$$\left( m_A H + (m_A + m_B)\frac{(R - R_0)^2}{4H} \right) \ddot{w}_A$$
$$+ \left( K_x H + K_z \frac{(R - R_0)^2}{4H} + K_\varphi \frac{1}{H} \right) w_A = M_p \tag{4.13}$$

When alternating voltage is applied to the piezoelectric electrodes:

$$U_p(t) = U_{p0} \sin \omega_p t \tag{4.14}$$

where $U_{d0}$ represents the driving voltage magnitude, and $\omega_d$ represents the driving voltage frequency. The simple harmonic driving moment generated by the flexural piezoelectric actuator is as follows:

$$M_p(t) = U_{p0} \left( \frac{c_{13}e_{33}}{c_{33}} - e_{31} \right) b_p \frac{h_b + h_p}{2} \sin \omega_p t = M_{p0} \sin \omega_p t \tag{4.15}$$

where $M_{p0}$ represents the magnitude of the harmonic driving moment $M_p(t)$.

Equation (4.15) is then substituted into Eq. (4.13), with the following equation derived:

$$m_d^* \ddot{w}_A + k_d^* w_A = F^* \sin \omega_p t \tag{4.16}$$

where $m_d^*$ represents the generalized mass of the resonator; $k_d^*$ represents the generalized stiffness the resonator; and $F^*$ represents the generalized load. Their expressions are as follows:

$$\begin{cases} m_d^* = m_A + (m_A + m_B)\frac{(R-R_0)^2}{4H^2} \\ F^* = \frac{M_{p0}}{H} = U_p(t)\frac{\Theta_{UM}}{H} \\ k_d^* = K_x + K_z\frac{(R-R_0)^2}{4H^2} + K_\varphi\frac{1}{H^2} \end{cases} \tag{4.17}$$

The resonant frequency of the single-DOF system in Eq. (4.16) can be evaluated by substituting the generalization mass $m_d^*$ and generalized stiffness $k_d^*$ of the resonator into the equation through the Rayleigh method, thus:

$$\omega_d = \sqrt{\frac{k_d^*}{m_d^*}} = \sqrt{\frac{4K_x H^2 + K_z(R - R_0)^2 + 4K_\varphi}{4m_A H^2 + (m_A + m_B)(R - R_0)^2}} \tag{4.18}$$

Equation (4.18) is then substituted into Eq. (4.16), deriving a kinetic equation for the system in drive mode, as follows:

$$\ddot{w}_A + \omega_d^2 w_A = \frac{F^*}{m_d^*}\sin\omega_p t \tag{4.19}$$

Considering the vibration damping effect of the resonator, the equivalent viscous damping ratio $\xi$ of the single-DOF system is introduced. Owing to the high complexity of the factors affecting the damping effect of the resonator, the damping ratio can usually be determined experimentally. There are about 5000–40,000 mechanical quality factors in a typical metal resonator, and the relevant damping ratio is about $10^{-4}$. Considering the influence of the damping ratio, the kinetic Eq. (4.19) of the system can be rewritten as:

$$\ddot{w}_A + 2\xi\omega_d w_A + \omega_d^2 w_A = \frac{F^*}{m_d^*}\sin\omega_p t \tag{4.20}$$

### 4.2.1 Steady-State Responses in Drive Mode

#### I.   Steady-state Solution

Equation (4.20) describes the motion state of the resonator under a simple harmonic driving force in damping conditions. This is a kinetic equation for a typical single-DOF system. According to the structural kinetic theory of single-DOF systems, the kinetic Eq. (4.20) of the single-DOF system has the following homogeneous solution:

$$\overline{w}_A(t) = (C_1 \cos\omega_c t + C_2 \sin\omega_c t)e^{-\xi\omega_d t} \tag{4.21}$$

where $\omega_c = \omega_d\sqrt{1 - \xi^2}$ represents a free vibration frequency for a damping system.

The particular solution of Eq. (4.20) can be written as follows:

$$w_A^*(t) = G_1 \cos \omega_d t + G_2 \sin \omega_d t \tag{4.22}$$

So,

$$
\begin{aligned}
w_A^*(t) &= \frac{F^*}{m_d^*} \frac{1}{\left(\omega_p^2 - \omega_d^2\right)^2 + 4\xi^2 \omega_d^2} \left[\left(\omega_p^2 - \omega_d^2\right) \sin \omega_p t - 2\xi \omega_d^2 \cos \omega_p t\right] \\
&= \frac{F^*}{k_d^*} \frac{1}{(1 - \upsilon_d^2)^2 + 4\xi^2 \upsilon_d^2} \left[\left(1 - \upsilon_d^2\right) \sin \omega_p t - 2\xi \upsilon_d \cos \omega_p t\right] \quad (4.23)
\end{aligned}
$$

where $\upsilon_d = \frac{\omega_p}{\omega_d}$ represents the frequency ratio in drive mode.

The general solution of the kinetic equation of the generalized single-DOF system in the drive mode of the resonator is the sum of the homogeneous solution and the particular solution, as follows:

$$
\begin{aligned}
w_A(t) &= \overline{w}_A(t) + w_A^*(t) \\
&= (C_1 \cos \omega_c t + C_2 \sin \omega_c t) e^{-\xi \omega_d t} \\
&\quad + \frac{F^*}{k_d^*} \frac{1}{(1 - \upsilon_d^2)^2 + 4\xi^2 \upsilon_d^2} \left[\left(1 - \upsilon_d^2\right) \sin \omega_p t - 2\xi \upsilon_d \cos \omega_p t\right] \quad (4.24)
\end{aligned}
$$

The first term to the right of the equals sign in Eq. (4.24) is a free vibrational term determined according to the initial conditions of the system. Owing to damping attenuation, the first term will gradually disappear after a given period of motion, so this is called the transient response or transition state of the system; the second term to the right of the equals sign is related to the generalized force. It vibrates at the frequency of the driving force and does not attenuate over time, so it is known as the steady-state response or forced vibration of the system.

The steady-state response of the resonator in drive mode can be written as follows:

$$
\begin{aligned}
w_A(t) &= \frac{F^*}{k_d^*} \frac{1}{\left(1 - \upsilon_d^2\right)^2 + 4\xi^2 \upsilon_d^2} \left[\left(1 - \upsilon_d^2\right) \sin \omega_p t - 2\xi \upsilon_d \cos \omega_p t\right] \\
&= \frac{F^*}{k_d^*} \frac{1}{\sqrt{\left(1 - \upsilon_d^2\right)^2 + 4\xi^2 \upsilon_d^2}} \\
&\quad \left[\frac{\left(1 - \upsilon_d^2\right)}{\sqrt{\left(1 - \upsilon_d^2\right)^2 + 4\xi^2 \upsilon_d^2}} \sin \omega_p t - \frac{2\xi \upsilon_d}{\sqrt{\left(1 - \upsilon_d^2\right)^2 + 4\xi^2 \upsilon_d^2}} \cos \omega_p t\right] \\
&= w_{A\_st} \eta_d \sin(\omega_p t - \beta_d) = w_{A\_d} \sin(\omega_p t - \beta_d) \quad (4.25)
\end{aligned}
$$

where $w_{A\_d}$ represents the magnitude of dynamic response; $w_{A\_st}$ represents the static displacement; $\eta_d$ represents the amplification coefficient of the dynamic displacement in drive mode; and $\beta_d$ represents the response lag phase angle of the dynamic displacement in drive mode. Their expressions are as follows:

$$\begin{cases} w_{A\_st} = \frac{F^*}{k_d^*} \\ w_{A\_d} = w_{A\_st}\eta_d \\ \eta_d = \frac{1}{\sqrt{(1-v_d^2)^2+4\xi^2 v_d^2}} \\ \beta_d = \arccos\left(\frac{1-v_d^2}{\sqrt{(1-v_d^2)^2+4\xi^2 v_d^2}}\right) \end{cases} \quad (4.26)$$

According to the above analysis, the steady-state response of the resonator in drive mode has the following characteristics:

(1) The steady-state response frequency of the resonator is equal to the frequency of the driving moment.
(2) The amplitude of the steady-state response of the resonator has nothing to do with the initial conditions and does not change over time. The amplitude is the product of the static amplitude and the dynamic amplification coefficient under the driving force.

II. **Amplitude-Frequency Characteristics of the Resonator**

Equation (4.26) can be used to produce the relation curve between the dynamic amplifying coefficient $\eta_d$ and the frequency ratio $v$ of the steady-state response of the resonator, i.e., the amplitude-frequency curve of the resonator. As the damping ratio of the resonator changes, its amplitude-frequency response curve will change, too. The amplitude-frequency response curve for high system damping is shown in Fig. 4.5a. Taking a common damping ratio $(\xi = 10^{-4})$ in CVG resonators as an example, the relevant amplitude-frequency response curve is shown in Fig. 4.5b.

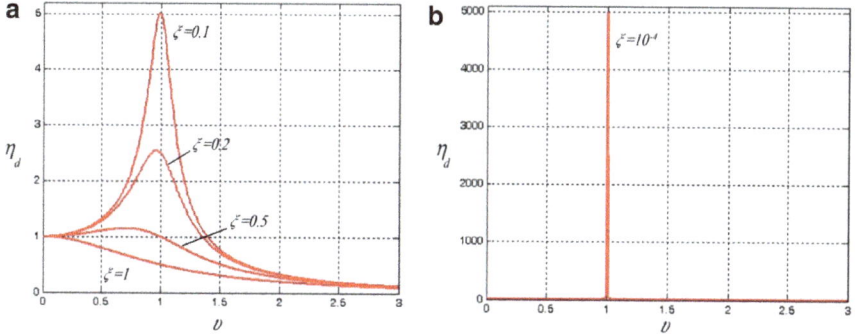

**Fig. 4.5** Amplitude-frequency characteristic curve of the resonator model. **a** System at high damping ratio; **b** system at ultra-low damping ratio

As seen in Eq. (4.26) and Fig. 4.5a, the dynamic response of any damping system reaches a maximum near the frequency ratio. At this time, the magnitude of system damping has a major influence on the amplitude of the system's steady-state responses; when the frequency ratio $\upsilon \ll 1$, the amplitude of the steady-state response of the system is equivalent to its static displacement. When the frequency ratio $\upsilon > 2$, there is basically no difference in the influence of the magnitude of system damping on the amplitude of the steady-state response of the system. When the frequency ratio is very high, the steady-state response of the system approaches zero.

As shown in Fig. 4.5, the CVG resonator has a very high number of mechanical quality factors and the system has an ultra-low damping ratio. When the frequency ratio $\upsilon = 1$ (the driving force frequency is consistent with a certain-order modal frequency, i.e., $\upsilon = \sqrt{1 - \xi^2} \approx 1$ during resonant vibrations, the dynamic amplification coefficient is highest. At this time, the dynamic response of the resonator reaches its peak. If the frequency ratio is beyond $\upsilon = 1$, the dynamic amplification coefficient drops sharply and can be ignored compared with the period of resonant vibrations.

III.    **Phase-Frequency Characteristics of the Resonator**

Equation (4.26) can be used to produce the relation curve between the dynamic displacement response lag phase angle $\beta$ and frequency ratio $\upsilon$ of the steady-state response of the resonator, i.e., the phase-frequency characteristic curve of the resonator (see Fig. 4.6). As the damping ratio of the resonator changes, its phase-frequency characteristic curve will change accordingly.

According to Eq. (4.26) and Fig. 4.6a, when the frequency ratio $\upsilon = \sqrt{1 - \xi^2}$, the dynamic response phase angle of any damping system is 90°, and this is an important feature of second-order system resonance. This method can be used to measure the natural frequency of a system. When the frequency ratio $\upsilon \ll 1$, the dynamic response phase angle of the system approaches 0°, and when the frequency ratio $\upsilon \gg 1$, the dynamic response phase angle of the system approaches 180°.

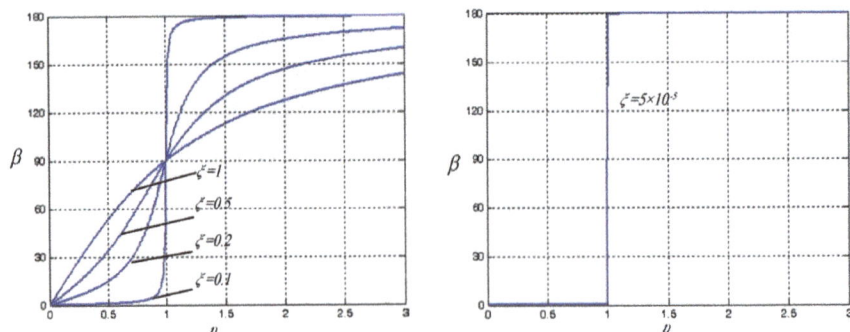

**Fig. 4.6**  Phase-frequency characteristic curve of the resonator model

As shown in Fig. 4.6b, a CVG resonator has a very high mechanical quality factor and the system has an ultra-low damping ratio. When the frequency ratio $\upsilon = 1$ (the driving force frequency is consistent with a certain-order modal frequency, i.e., $\upsilon = \sqrt{1 - \xi^2} \approx 1$ during resonant vibrations), the dynamic response phase angle of the system is equal to 90°; when $\upsilon < 1$, the dynamic response phase angle of the system rapidly drops to 0°; when $\upsilon > 1$, the dynamic response phase angle of the system rapidly rises to 180°. This shows that in the drive control of a gyroscope, the dynamic response phase angle of its resonator can be controlled at 90°, so that the resonator can remain in a state of constant resonance.

According to the results of the above analysis, for resonators with high mechanical quality factors, their systems will have ultra-low damping ratios. When the driving force frequency of the resonator is consistent with the frequency of its working modes, the system has the highest dynamic response to the shapes of its working modes, but a negligible dynamic response to other mode shapes. The higher the mechanical quality factors of the resonator are, the more prominent this phenomenon becomes. Therefore, high dynamic response accuracy can be obtained by performing a steady-state response analysis on the driving force in a resonator's working modes. At this time, Eq. (4.25) is expressed as a steady-state response in a second-order system under simple harmonic excitation. In general, a second-order system can be used to approximately represent the vibrations of a CVG resonator. Equation (4.6) can be rewritten as:

$$
\begin{bmatrix} u \\ v \\ w \end{bmatrix} = \sum_{i=1}^{\infty} \begin{bmatrix} \phi_{u\_i} Y_{u\_i} \\ \phi_{v\_i} Y_{v\_i} \\ \phi_{w\_i} Y_{w\_i} \end{bmatrix} \approx \begin{bmatrix} \phi_{u\_N} Y_{u\_N} \\ \phi_{v\_N} Y_{v\_N} \\ \phi_{w\_N} Y_{w\_N} \end{bmatrix} \tag{4.27}
$$

where $N$ represents the order of the working modes of the resonator.

Therefore, according to the above kinetic analysis of resonators, when the driving frequency of the flexural piezoelectric actuator is equal to the drive mode frequency of the resonator, i.e., when $\omega_p = \omega_d$, the drive mode of the resonator is excited and the amplitude of its dynamic displacement reaches a maximum as a steady-state response, therefore:

$$
w_A(t) = \frac{F^*}{k_d^*} \frac{1}{2\xi} \sin\left(\omega_d t - \frac{\pi}{2}\right) \tag{4.28}
$$

At this time, the dynamic amplification coefficient of the resonator's dynamic response is $1/2\xi$, and the response lag phase angle is 90°.

## IV.  Mode Shape Distribution Function of Drive Mode

After the derivation of a generalized displacement function for the dynamic response of the resonator, a vibration displacement distribution function can be established for the steady-state responses of each point on the resonator in drive mode based on the mode shape function of the resonator in its working modes.

The vibration displacement distribution function of the shell wall of the resonator in drive mode can be written as follows:

$$\begin{cases} u_{c\_d}(r, \theta, z, t) = u_{A\_d} \cos 2\theta \sin \omega_p t = -w_{A\_d} \frac{R-R_0}{2H} \cos 2\theta \sin(\omega_p t - \beta_d) \\ v_{c\_d}(r, \theta, z, t) = v_{A\_d} \frac{z}{H} \sin 2\theta \sin \omega_p t = \frac{1}{2} w_{A\_d} \frac{z}{H} \sin 2\theta \sin(\omega_p t - \beta_d) \\ w_{c\_d}(r, \theta, z, t) = w_{A\_d} \frac{z}{H} \cos 2\theta \sin(\omega_p t - \beta_d) \end{cases} \tag{4.29}$$

The vibration displacement distribution function of the bottom of the resonator in drive mode can be written as follows:

$$\begin{cases} u_{b\_d}(r, \theta, z, t) = u_{A\_d} \left(\frac{r-R_0}{R-R_0}\right)^2 \cos 2\theta \sin(\omega_p t - \beta_d) \\ \qquad\qquad = -w_{A\_d} \frac{R-R_0}{2H} \left(\frac{r-R_0}{R-R_0}\right)^2 \cos 2\theta \sin(\omega_p t - \beta_d) \\ v_{b\_d}(r, \theta, z, t) = 0 \\ w_{b\_d}(r, \theta, z, t) = 0 \end{cases} \tag{4.30}$$

### 4.2.2   Detection Signals of Drive Modes

Based on the basic working principles of gyroscopes, the piezoelectric electrodes used for detection in a resonator's drive mode are distributed at the point on the bottom of the resonator where $\theta = \pi/2$ and $\theta = 3\pi/2$. Taking the piezoelectric electrode at $\theta = \pi/2$ as an example, according to Eq. (4.30), its axial displacement function can be expressed in the local coordinate system of the flexural piezoelectric sensor as:

$$u_{b\_d}(x, t) = -u_{A\_d} \left(\frac{x - R_0}{R - R_0}\right)^2 \sin(\omega_p t - \beta_d) \tag{4.31}$$

According to the principle of flexural piezoelectric sensors, the piezoelectric electrodes are subjected to flexural vibrations with the bottom of the resonator in the drive mode of the CVG. Its major internal strain is an $x$-axial normal strain, and the $x$-axial normal strain in the neutral plane is related to the deflection curve at the bottom of the resonator, as shown in Fig. 4.7.

According to Euler–Bernoulli beam theory, after a piezoelectric electrode deflects from the bottom of the resonator at the value of $u$, its $x$-axial normal strain in the middle plane is equal to:

$$S(x, t) = \frac{h_b + h_p}{2R_\varphi(x, t)} = \frac{h_b + h_p}{2} \chi(x, t) = \frac{h_b + h_p}{2} \frac{u''}{\sqrt{(1 + u'^2)^3}} \tag{4.32}$$

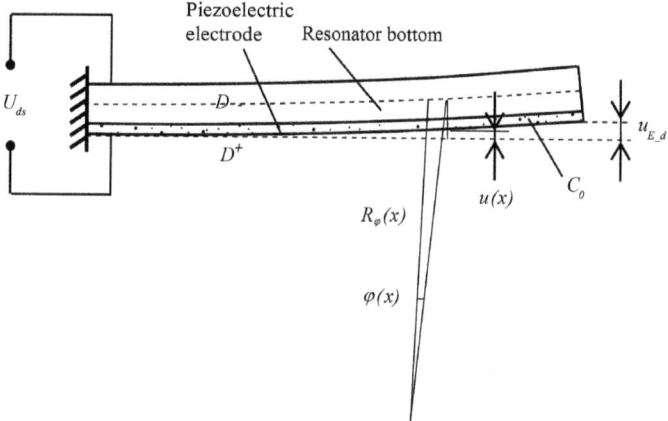

**Fig. 4.7** Strain and signal of the flexural piezoelectric sensor in drive mode

where $R_\varphi(x)$ represents the curvature radius of the flexural piezoelectric sensor; $\chi(x)$ represents the curvature; and $u'' = \frac{\partial^2 u}{\partial x^2}$ represents the deflection of the flexural piezoelectric sensor along the $x$-axis. Because the deflection of the flexural piezoelectric sensor is much smaller than its length dimension, $u'$ is negligible compared to 1, and Eq. (4.32) can be rewritten as:

$$S(x, t) = \frac{h_b + h_p}{2} \frac{\partial^2 u}{\partial x^2} \tag{4.33}$$

Equation (4.31) is then substituted into Eq. (4.32), with the following equation derived:

$$S(x, t) = -u_{A\_d} \frac{h_b + h_p}{(R - R_0)^2} \sin(\omega_p t - \beta_d) \tag{4.34}$$

Equation (4.33) is then substituted into Eq. (2.20), allowing for the output detection voltage $U_{ds}$ on the piezoelectric electrode of the flexural piezoelectric sensor in drive mode, as follows:

$$U_{ds}(t) = \frac{h_p e_{31} \int_0^{l_p} S(x) dx}{\varepsilon_{33} l_p} = -\frac{e_{31} h_p \int_0^{R-R_0} u_{A\_d}(h_b + h_p) \sin(\omega_p t - \beta_d) dx}{\varepsilon_{33}(R - R_0)^3}$$

$$= w_{A\_d} \frac{e_{31} h_p (h_b + h_p)}{2H\varepsilon_{33}(R - R_0)} \sin(\omega_p t - \beta_d) \tag{4.35}$$

Equations (4.25) and (4.26) are then substituted into Eq. (4.35), allowing for the following statement to be derived:

$$U_{ds} = \frac{U_p(t)}{k_d^*}\left(\frac{c_{13}e_{33}}{c_{33}} - e_{31}\right)b_p \frac{h_b + h_p}{2H\sqrt{\left(1 - v_d^2\right)^2 + 4\xi^2 v_d^2}}$$

$$\frac{e_{31}h_p(h_b + h_p)}{2H\varepsilon_{33}(R - R_0)^2}\sin(\omega_p t - \beta_d)$$

$$= \frac{U_p(t)}{k_d^*}\frac{1}{\sqrt{\left(1 - v_d^2\right)^2 + 4\xi^2 v_d^2}}K_G K_P \sin(\omega_p t - \beta_d) \qquad (4.36)$$

where $K_G$ is the resonator's geometric parameter influence coefficient of the resonator and $K_P$ is the resonator's physical parameter influence coefficient. Their respective expressions are as follows:

$$\begin{cases} K_G = \frac{b_p h_p (h_b + h_p)^2}{4H^2(R - R_0)} \\ K_P = \left(\frac{c_{13}e_{33}}{c_{33}} - e_{31}\right)\frac{e_{31}}{\varepsilon_{33}} \end{cases} \qquad (4.37)$$

Equation (4.24) shows that, ideally, output signal $U_{ds}$ from a flexural piezoelectric sensor in drive mode has the same frequency as the input signal $U_d$ of the gyroscope's driving force, and their phase difference is $\pi/2$.

## 4.3  Coriolis Force in Resonators

### 4.3.1  Velocity of Vibration and Coriolis Force

According to the results of the analysis on the steady-state response of the resonator in drive mode, the resonator is subjected to forced vibrations under the excitation of the flexural moment of the flexural piezoelectric actuator. When the exciting voltage has the same frequency as the natural frequency of the resonator in its working modes, the resonator starts working in drive mode and its amplitude reaches a maximum. The vibration displacement function of its steady-state response at each point is shown in Eqs. (4.38) and (4.39). The displacement at each point is differentiated with respect to time, deriving a vibration velocity expression for the shell wall and the bottom of the resonator in drive mode, and these figures are shown in the following equations:

$$\begin{cases} \dot{u}_{c\_d}(\theta, z, t) = \frac{du_{c\_d}(\theta,z,t)}{dt} = \dot{u}_{A\_d}\cos 2\theta \cos(\omega_p t - \beta_d) \\ \dot{v}_{c\_d}(\theta, z, t) = \frac{dv_{c\_d}(\theta,z,t)}{dt} = \dot{v}_{A\_d}\frac{z}{H}\sin 2\theta \cos(\omega_p t - \beta_d) \\ \dot{w}_{c\_d}(\theta, z, t) = \frac{dw_{c\_d}(\theta,z,t)}{dt} = \dot{w}_{A\_d}\frac{z}{H}\cos 2\theta \cos(\omega_p t - \beta_d) \end{cases} \qquad (4.38)$$

$$\begin{cases} \dot{u}_{b\_d}(\theta, r, t) = \frac{du_{b\_a}(\theta,r,t)}{dt} = \dot{u}_{A\_d}\left(\frac{r - R_0}{R - R_0}\right)^2 \cos 2\theta \cos(\omega_p t - \beta_d) \\ \dot{v}_{b\_d}(\theta, r, t) = 0, \dot{w}_{b\_d}(\theta, r, t) = 0 \end{cases} \qquad (4.39)$$

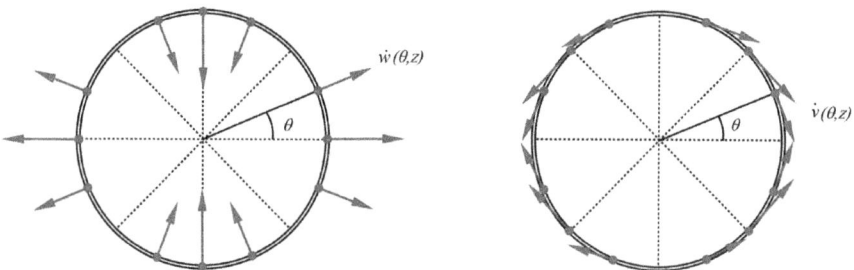

**Fig. 4.8** Distribution of the vibration velocity at each point on the resonator shell wall in the *xoy* plane

where $[\dot{u}_{A\_d}, \dot{v}_{A\_d}, \dot{w}_{A\_d}]$ represents the maximum vibration velocity of the resonator in drive mode:

$$\begin{cases} \dot{u}_{A\_d} = u_{A\_d}\omega_p \\ \dot{v}_{A\_d} = v_{A\_d}\omega_p \\ \dot{w}_{A\_d} = w_{A\_d}\omega_p \end{cases} \tag{4.40}$$

Considering that CVGs are used to detect the $z$-axial angular velocity, according to the formation mechanism of the Coriolis force [2], Coriolis force is only generated when the vibration velocity of the resonator is orthogonal to the detection axis, and Coriolis force will not be generated when the vibration velocity is parallel to the detection axis. Therefore, the axial vibration velocity at the points on the resonator does not generate Coriolis force. Therefore, when the Coriolis force in the resonator is analyzed, only the tangential vibration velocity and radial vibration velocity at each point on the resonator's shell wall are considered. The distribution of the tangential vibration velocity and radial vibration velocity at each point on the resonator shell wall in the *xoy* plane are shown in Fig. 4.8.

Based on the above analysis, when the gyroscope rotates around the $z$-axis, the radial vibration velocity and tangential vibration velocity at each point on the resonator shell wall will generate Coriolis force. According to the definition of Coriolis force, i.e., $F_c = -2\,m \times v$, under the action of the $z$-axial angular velocity $\Omega$, the infinitesimal Coriolis force on the shell wall of the resonator generated by the radial vibration velocity and tangential vibration velocity is as follows:

$$\begin{cases} f_{cv}(\theta, z, t) = \dot{w}_{c\_d}(\theta, z, t)\Omega dm \cos(\omega_p t - \beta_d) \\ f_{cw}(\theta, z, t) = \dot{v}_{c\_d}(\theta, z, t)\Omega dm \cos(\omega_p t - \beta_d) \end{cases} \tag{4.41}$$

where $dm$ represents the infinitesimal mass at each point on the resonator; $f_{cv}(\theta, z)$ represents the tangential Coriolis force generated by the radial vibration velocity at each point on the resonator; and $f_{cv}(\theta, z)$ represents the radial Coriolis force by the tangential vibration velocity at each point on the resonator.

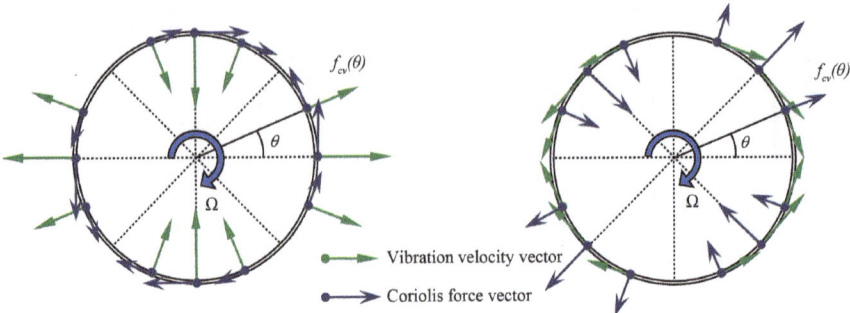

**Fig. 4.9** Distribution of the Coriolis force at each point on the shell wall of the resonator in the *xoy* plane

Equation (4.38) is then substituted into Eq. (4.41), allowing for the calculation of the infinitesimal Coriolis force as follows:

$$\begin{cases} f_{cv}(\theta, z) = \dot{w}_{A\_d}\frac{z}{H}\Omega dm \cos 2\theta = \dot{w}_{A\_d}\frac{z}{H}\Omega(\rho t_c dz R d\theta) \cos 2\theta \\ \quad = \left(\dot{w}_{A\_d}\frac{z}{H}\Omega \rho t_c R dz\right) \cos 2\theta d\theta = p_{cv}(z) \cos 2\theta d\theta \\ f_{cw}(\theta, z) = \dot{v}_{A\_d}\frac{z}{H}\Omega dm \sin 2\theta = \dot{v}_{A\_d}\frac{z}{H}\Omega(\rho t_c dz R d\theta) \sin 2\theta \\ \quad \left(\dot{v}_{A\_d}\frac{z}{H}\Omega \rho t_c R dz\right) \sin 2\theta d\theta = p_{cw}(z) \sin 2\theta d\theta \end{cases} \quad (4.42)$$

where $p_{cv}(z)$ and $p_{cw}(z)$ respectively represent the load density of the tangential Coriolis force and the load density of the radial Coriolis force. Their expressions are as follows:

$$\begin{cases} p_{cv}(z) = \dot{w}_{A\_d}\frac{z}{H}\Omega \rho t_c R dz \\ p_{cw}(z) = \dot{v}_{A\_d}\frac{z}{H}\Omega \rho t_c R dz \end{cases} \quad (4.43)$$

The distribution of the radial Coriolis force and tangential Coriolis force at each point on the shell wall of the resonator in the *xoy* plane can be determined according to Eq. (4.42), as shown in Fig. 4.9.

### 4.3.2 Coriolis Force and Equivalent Moment of Force

The shell wall of the resonator is discretized into a number of rings along the *z*-axis to analyze the deformation of the ring at height $dz$ as caused by the Coriolis force. Considering the Coriolis force on the ring and the structural symmetry of the ring, a micro-segment of the ring is taken from the $-45°$ to $+45°$ range circumferentially for the purposes of force analysis. As shown in Fig. 4.10, let the normal stress on the circumferential section be $N(\theta)$, the shear stress be $Q(\theta)$, and the inner flexural moment be $M(\theta)$. The infinitesimal tangential Coriolis force and radial Coriolis force are regarded as external forces.

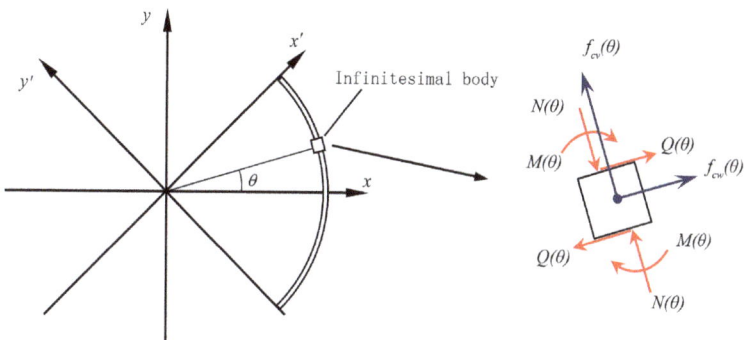

**Fig. 4.10** Schematic diagram of the force borne by a micro-segment of a discrete ring on the shell wall of the resonator

After decomposition along the $x'$ axis and $y'$ axis, the tangential Coriolis force $f_{cv}(\theta, z)$ on the ring is integrated as follows:

$$
\begin{cases}
N_{xv} = \int_\theta^{\frac{\pi}{4}} f_{cv}(\theta) \cos\left(\frac{\pi}{4} - \theta\right) d\theta = \int_\theta^{\frac{\pi}{4}} p_{cv}(z) \cos 2\theta \sin\left(\frac{\pi}{4} - \theta\right) d\theta \\
\quad = p_{cv}(z)\left[\sin\left(\frac{\pi}{4} - \theta\right) - \frac{1}{3}\sin\left(\frac{3\pi}{4} - 3\theta\right)\right] \\
N_{yv} = -\int_{-\frac{\pi}{4}}^\theta f_{cv}(\theta) \cos\left(\frac{\pi}{4} + \theta\right) d\theta = -\int_{-\frac{\pi}{4}}^\theta p_{cv}(z) \cos 2\theta \sin\left(\frac{\pi}{4} + \theta\right) d\theta \\
\quad = p_{cv}(z)\left[\cos\left(\frac{\pi}{4} - \theta\right) + \frac{1}{3}\cos\left(\frac{3\pi}{4} - 3\theta\right)\right]
\end{cases}
\tag{4.44}
$$

After decomposition along the $x'$ axis and $y'$ axis, the radial Coriolis force $f_{cv}(\theta, z)$ on the ring is integrated as follows:

$$
\begin{cases}
N_{xw} = \int_\theta^{\frac{\pi}{4}} f_{cw}(\theta) \cos\left(\frac{\pi}{4} - \theta\right) d\theta = \int_\theta^{\frac{\pi}{4}} p_{cw}(z) \sin 2\theta \cos\left(\frac{\pi}{4} - \theta\right) d\theta \\
\quad = p_{cw}(z)\left[\sin\left(\frac{\pi}{4} - \theta\right) + \frac{1}{3}\sin\left(\frac{3\pi}{4} - 3\theta\right)\right] \\
N_{yw} = -\int_{-\frac{\pi}{4}}^\theta f_{cw}(\theta) \cos\left(\frac{\pi}{4} + \theta\right) d\theta = -\int_{-\frac{\pi}{4}}^\theta p_{cw}(z) \sin 2\theta \cos\left(\frac{\pi}{4} + \theta\right) d\theta \\
\quad = p_{cw}(z)\left[\cos\left(\frac{\pi}{4} - \theta\right) - \frac{1}{3}\cos\left(\frac{3\pi}{4} - 3\theta\right)\right]
\end{cases}
\tag{4.45}
$$

An $x'$-axial force equilibrium equation is established for the ring segment at $\theta$-45° and a $y'$-axial force equilibrium equation is established for the ring segment at $-45°$-$\theta$, as follows:

$$
\begin{cases}
N_{xv} + N_{xw} = Q(\theta) \cos\left(\frac{\pi}{4} - \theta\right) - N(\theta) \sin\left(\frac{\pi}{4} - \theta\right) \\
N_{yv} + N_{yw} = N(\theta) \cos\left(\frac{\pi}{4} - \theta\right) + Q(\theta) \sin\left(\frac{\pi}{4} - \theta\right)
\end{cases}
\tag{4.46}
$$

Equations (4.44) and (4.45) is substituted into Eq. (4.46), resulting in the solution:

$$\begin{cases} Q(\theta) = \left(\frac{2}{3}p_{cv}(z) + \frac{4}{3}p_{cw}(z)\right)\cos 2\theta \\ N(\theta) = \left(\frac{4}{3}p_{cv}(z) + \frac{2}{3}p_{cw}(z)\right)\sin 2\theta \end{cases} \tag{4.47}$$

According to the equation for the infinitesimal moment equilibrium:

$$\Delta M = Q(\theta)R\Delta\theta \tag{4.48}$$

The following infinitesimal moment differentiation equation is then derived:

$$\frac{\partial M}{\partial \theta} = Q(\theta)R = R\left(\frac{2}{3}p_{cv}(z) + \frac{4}{3}p_{cw}(z)\right)\cos 2\theta \tag{4.49}$$

The solution is below:

$$M(\theta) = R\left(\frac{1}{3}p_{cv}(z) + \frac{2}{3}p_{cw}(z)\right)\sin 2\theta \tag{4.50}$$

Therefore, the displacement of the ring under the Coriolis force is evaluated after obtaining the internal normal stress, shear stress, and internal moment of the ring. The results are below:

$$\frac{\partial^2 w}{\partial \theta^2} + w = -\frac{M(\theta)R^2}{EI} = -\left(\frac{1}{3}p_{cv}(z) + \frac{2}{3}p_{cw}(z)\right)\frac{R^3}{EI}\sin 2\theta \tag{4.51}$$

where $I = \frac{t_c^3 dz}{12}$ represents the moment of inertia of the section of the ring at height $dz$. The differential Eq. (4.24) can then be solved, deriving the radial displacement function expression of the ring under the Coriolis force, as follows:

$$w = \left(\frac{1}{9}p_{cv}(z) + \frac{2}{9}p_{cw}(z)\right)\frac{R^3}{EI}\sin 2\theta = w_{c0}\sin 2\theta \tag{4.52}$$

The radial displacement function expression (4.24) is then substituted into Eq. (3.16), deriving the tangential displacement function expression of the ring under the Coriolis force, as follows:

$$v = \left(\frac{1}{18}p_{cv}(z) + \frac{1}{9}p_{cw}(z)\right)\frac{R^3}{EI}\cos 2\theta = v_{c0}\cos 2\theta \tag{4.53}$$

According to the analysis of radial ring stiffness given in Chap. 3, the displacement function of the ring under the Coriolis force shows the same rule of circumferential distribution as the displacement function of the ring under the concentrated force from two symmetric points. Therefore, the Coriolis force on the ring can be treated as equivalent to a concentrated force applied from two symmetric points; i.e., the

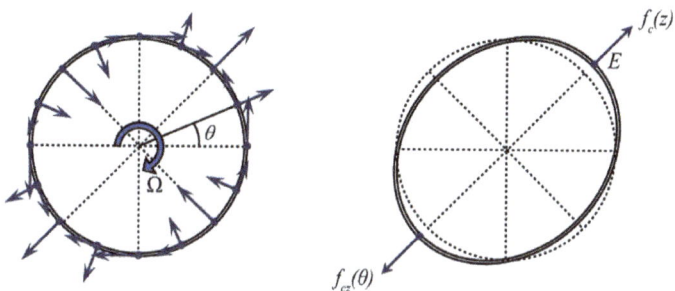

**Fig. 4.11** Distribution of the Coriolis force on the shell wall of the resonator as a discrete ring and its equivalent force

equivalent force only does work at $\theta = \pi/4$ and $\theta = 5\pi/4$ and deforms the ring in the pattern shown in Fig. 4.11.

The equivalent of the Coriolis force can be obtained according to Eq. (4.58) and Eq. (4.26), as follows:

$$f_c(z) = w_{c0} \cdot dk_w = \left(\frac{1}{9}p_{cv}(z) + \frac{2}{9}p_{cw}(z)\right)\frac{R^3}{EI}\frac{3\pi\,Et_c^3 dz}{8R^3}$$

$$= \frac{\pi}{2}p_{cv}(z) + \pi p_{cw}(z) \tag{4.54}$$

The above derivation is based on the axial discretization of the shell wall of the resonator into a number of rings. Equation (4.43) shows that the rings at different heights $dz$ have different geometric parameters and vibration velocities. Equation (4.43) is then substituted into Eq. (4.54), with the following statement derived:

$$f_c(z) = \frac{\pi}{2}\frac{z}{H}\Omega\rho Rt\left(\dot{w}_{A\_d} + 2\dot{v}_{A\_d}\right)dz \tag{4.55}$$

Therefore, the distribution of $f_c(z)$, which is the equivalent of the Coriolis force on the ring at height $dz$ as described in Eq. (4.55), can be obtained at the point along the generatrix inside the shell wall of the resonator in the $x'oz$ plane, as shown in Fig. 4.12a.

According to the distribution of the equivalent Coriolis force along the generatrix inside the shell wall of the resonator, as well as Eq. (4.55), the equivalent Coriolis force on the shell wall of the resonator can generate a moment at the lower endpoint $B$ of the generatrix. That is, an equivalent moment of force is generated at Point $B$ in the resonator's lumped parameter model, as shown in Fig. 4.12b. The equivalent moment of Coriolis force generated in the lumped parameter model of the resonator can be calculated according to Eq. (4.55), as shown below:

**Fig. 4.12** The equivalent
force distribution on the
resonator shell wall and the
equivalent moment of force.
**a** Equivalent force
distribution; **b** equivalent
moment of force

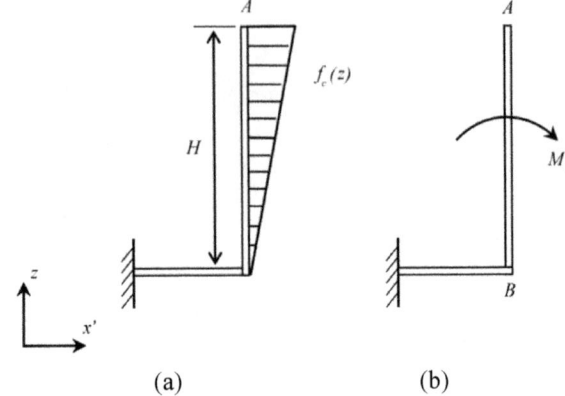

(a)                                            (b)

$$M_c = \int\limits_0^H \left( \frac{\pi}{2} \frac{z}{H} \Omega\rho R t_2 (\dot{w}_{A\_d} + 2\dot{v}_{A\_d}) \right) \cdot z\, dz$$

$$= \frac{\pi \Omega\rho R (\dot{w}_{A\_d} + 2\dot{v}_{A\_d}) t H^2}{6} \tag{4.56}$$

According to the definition of Coriolis force and Eq. (4.56), the Coriolis force on
the resonator is at the same frequency and phase as the steady-state response of the
resonator in drive mode, so the equivalent harmonic Coriolis moment is:

$$M_c(t) = \Omega\rho (\dot{w}_{A\_d} + 2\dot{v}_{A\_d}) \frac{\pi R t H^2}{6} \cos(\omega_p t - \beta_d)$$

$$= M_{c0} \sin\left(\omega_p t - \beta_d + \frac{\pi}{2}\right) \tag{4.57}$$

where $\omega_d$ represents the frequency of the equivalent harmonic Coriolis moment; and
$M_{c0}$ represents the amplitude of the equivalent harmonic Coriolis moment, relative
to the input angular velocity of the gyroscope. Equations (4.40) and (4.55) is then
substituted into Eq. (4.57), deriving the following expression:

$$M_{c0} = \Omega\rho \left( \omega_p \frac{F^*}{k_d^*} \frac{2}{\sqrt{(1 - v_d^2)^2 + 4\xi^2 v_d^2}} \right) \frac{\pi R t H^2}{6}$$

$$= \Omega \dot{w}_A \Theta_{\Omega M} \tag{4.58}$$

where $\Theta_{\Omega M} = \frac{\pi\rho R t H^2}{3}$ is a coefficient for the conversion of the angular velocity of
the resonator into the Coriolis moment.

## 4.4 Responses in Resonator Sense Mode

The Coriolis force on the CVG resonator is equivalent to the action of the moment. This equivalent moment can be regarded as a driving force at the same frequency and phase as the vibration velocity of the resonator in drive mode. This driving force enables the resonator to generate sense mode while it is in the appropriate position. Based on the symmetry of the resonator structure, using a resonator dynamics model, the present section analyzes the steady-state response of the resonator under the action of the equivalent moment and calculates the signals output from the piezoelectric electrodes in the resonator.

### 4.4.1 Steady-State Responses in Sense Mode

Based on the analysis of the Coriolis force on the resonator, the Coriolis moment acts on the resonator in the same way that the flexural piezoelectric driving moment acts on the resonator. From the structural symmetry of the resonator, the equivalent Coriolis moment can excite the sense mode of the resonator in this force state, and the resonator can also be equivalent to the equivalent lumped parameter model shown in Fig. 4.13.

Two lumped mass elements and three concentration stiffness elements in the equivalent lumped parameter model in Fig. 4.13 are identical to those in the equivalent lumped parameter model of the resonator in drive mode. $M_c$ represents the Coriolis

**Fig. 4.13** Equivalent lumped parameter model of the Coriolis moment and resonator

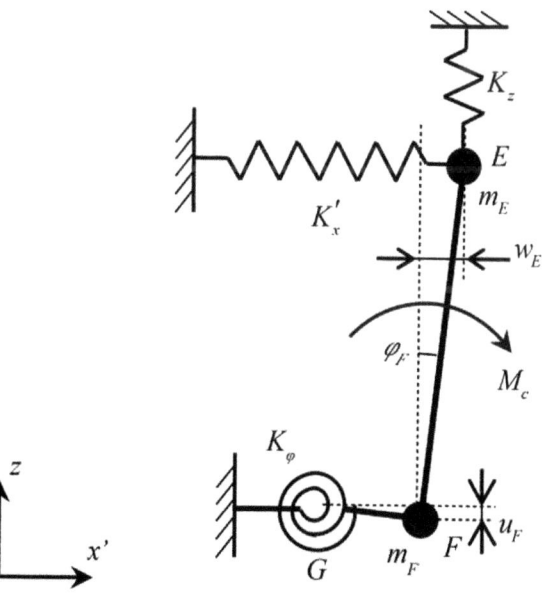

moment, and the generalized displacement is the radial displacement $w_E$ at Point $E$ on the shell wall top of the resonator (in the $x'oz$ plane).

The following is the kinetic equation of the equivalent lumped parameter model of a resonator under the action of the Coriolis moment:

$$
\left( m_E H + (m_E + m_F) \frac{(R - R_0)^2}{4H} \right) \ddot{w}_E
$$
$$
+ \left( K'_x H + K_z \frac{(R - R_0)^2}{4H} + \frac{E I_\theta}{(R - R_0)H} \right) w_E = M_c \tag{4.59}
$$

Based on the structural symmetry of the resonator, under ideal conditions, as given in Eq. (4.60):

$$
m_E = m_A, \; m_F = m_B, \; K'_x = K_x \tag{4.60}
$$

Considering the effect of the damping ratio, the dynamic Eq. (4.59) of the system in sense mode can be rewritten as:

$$
\ddot{w}_E + 2\xi \omega_s w_E + \omega_s^2 w_E = \frac{F_c^*}{m_s^*} \sin\left( \omega_p t - \beta_d + \frac{\pi}{2} \right) \tag{4.61}
$$

where $F_c^* = M_{c0}$ represents the generalized force in the system; $m_s^*$, frequency $\omega_s$, and damping $\xi$ are the same as those of the equivalent lumped parameter model of the system in drive mode. Along the same rationale, the general solution of Differential Eq. (4.61) is as follows:

$$
w_E(t) = (C_1 \cos \omega_c t + C_2 \sin \omega_c t) e^{-\xi \omega_s t}
$$
$$
+ \frac{F_c^*}{k_s^*} \frac{1}{\sqrt{(1 - v_s^2)^2 + 4\xi^2 v_s^2}} \left[ (1 - v_s^2) \sin\left( \omega_p t - \beta_d + \frac{\pi}{2} \right) - 2\xi v_s \cos \omega_d t \right] \tag{4.62}
$$

where $v_s = \frac{\omega_p}{\omega_s}$ represents the frequency ratio of sense mode.

The steady-state response of the resonator in sense mode can be written as:

$$
w_E(t) = \frac{F_c^*}{k_s^*} \frac{1}{\sqrt{(1 - v_s^2)^2 + 4\xi^2 v_s^2}} \left[ (1 - v_s^2) \sin\left( \omega_p t - \beta_d + \frac{\pi}{2} \right) - 2\xi v_s \cos \omega_d t \right]
$$
$$
= w_{E\_st} \eta_s \sin\left( \omega_p t - \beta_d + \frac{\pi}{2} - \beta_s \right) = w_{E\_s} \sin\left( \omega_p t - \beta_d + \frac{\pi}{2} - \beta_s \right) \tag{4.63}
$$

where $w_{E\_s}$ represents the dynamic displacement amplitude in sense mode; $w_{E\_st}$ represents the static displacement relative to the generalized force in the system; $\eta_s$ represents the dynamic displacement amplification coefficient in sense mode; and $\beta_s$ represents the response lag phase angle of the dynamic displacement in sense mode.

The above quantities are expressed as follows:

$$\begin{cases} w_{E\_s} = w_{E\_st} \eta_s \\ w_{E\_st} = \frac{F_c^*}{k_s^*} \\ \eta_s = \frac{1}{\sqrt{(1-v_s^2)^2+4\xi^2 v_s^2}} \\ \beta_s = \arccos\left(\frac{1-v_s^2}{\sqrt{(1-v_s^2)^2+4\xi^2 v_s^2}}\right) \end{cases} \quad (4.64)$$

From the above analysis, the dynamic characteristics of the resonator in sense mode excited by the Coriolis force can be shown to be basically equivalent to the dynamic characteristics of the resonator in drive mode excited by the flexural piezoelectric actuator.

According to the above dynamic analysis of the resonator in sense mode, when the harmonic Coriolis force is at the same frequency as the resonator's sense mode, i.e., when $\omega_p = \omega_s$, the sense mode of the resonator will be excited, and at the same time, the dynamic displacement of the resonator will reach maximum amplitude as a steady-state response, therefore:

$$w_E(t) = \frac{F_c^*}{k^*} \frac{1}{2\xi} \sin\left(\omega_p t - \frac{\pi}{2}\right) \quad (4.65)$$

At this time, the dynamic response amplification coefficient of the resonator is $1/2\xi$,, while the response lag phase angle is 90°.

After a generalized displacement function is derived for the dynamic response of the resonator in sense mode according to Eq. (4.65), a vibration displacement distribution function can be derived for the dynamic response of each point on the resonator in sense mode according to the mode shape function of the resonator in its working modes. The vibration displacement distribution function of the shell wall of the resonator in sense mode is as follows:

$$\begin{cases} u_{c\_s}(r, \theta, z, t) = u_{E\_s} \sin 2\theta \sin \omega_d t = w_{E\_s} \frac{R-R_0}{2H} \sin 2\theta \sin(\omega_p t - \beta_d + \frac{\pi}{2} - \beta_s) \\ v_{c\_s}(r, \theta, z, t) = -v_{E\_s} \frac{z}{H} \cos 2\theta \sin \omega_d t = -\frac{1}{2} w_{E\_s} \frac{z}{H} \cos 2\theta \sin(\omega_p t - \beta_d + \frac{\pi}{2} - \beta_s) \\ w_{c\_s}(r, \theta, z, t) = w_{E\_s} \frac{z}{H} \sin 2\theta \sin(\omega_p t - \beta_d + \frac{\pi}{2} - \beta_s) \end{cases} \quad (4.66)$$

The vibration displacement distribution function of the bottom of the resonator in sense mode is as follows:

$$\begin{cases} u_{b\_s}(r, \theta, z, t) = u_{E\_s} \left(\frac{r-R_0}{R-R_0}\right)^2 \sin 2\theta \sin(\omega_p t - \beta_d + \frac{\pi}{2} - \beta_s) \\ \qquad = -w_{E\_s} \frac{R-R_0}{2H} \left(\frac{r-R_0}{R-R_0}\right)^2 \sin 2\theta \sin(\omega_p t - \beta_d + \frac{\pi}{2} - \beta_s) \\ v_{b\_d}(r, \theta, z, t) = 0 \\ w_{b\_d}(r, \theta, z, t) = 0 \end{cases} \quad (4.67)$$

### 4.4.2  Detection Signals in Sense Mode

As can be seen from the basic working principles of gyroscopes, the piezoelectric electrodes used for detection in the resonator's sense mode are distributed at the point on the bottom of the resonator where $\theta = \pi/4$ and $\theta = 5\pi/4$. Taking the piezoelectric electrode at $\theta = \pi/4$ as an example, according to Eq. (4.67), its axial displacement function can be expressed in the local coordinate system of the flexural piezoelectric sensor as:

$$u_{b\_s}(x, t) = u_{E\_s}\left(\frac{x - R_0}{R - R_0}\right)^2 \sin\left(\omega_p t - \beta_d + \frac{\pi}{2} - \beta_s\right) \qquad (4.68)$$

According to the working principle of flexural piezoelectric actuators, in a CVG's sense mode, the piezoelectric electrodes are subjected to flexural vibrations with the bottom of the resonator. Its major internal strain is $x$-axial normal strain, and the $x$-axial normal strain in the neutral plane is related to the deflection curve at the bottom of the resonator, as shown in Fig. 4.14.

As in the case of vibration detection in drive mode, after a piezoelectric electrode deflects from the bottom of the resonator at the value of $u$, its $x$-axial normal strain in the middle plane is equal to:

$$S(x, t) = \frac{h_b + h_p}{2} \frac{\partial^2 u}{\partial x^2} \qquad (4.69)$$

Equation (4.68) is then substituted into Eq. (4.69), with the following equation derived:

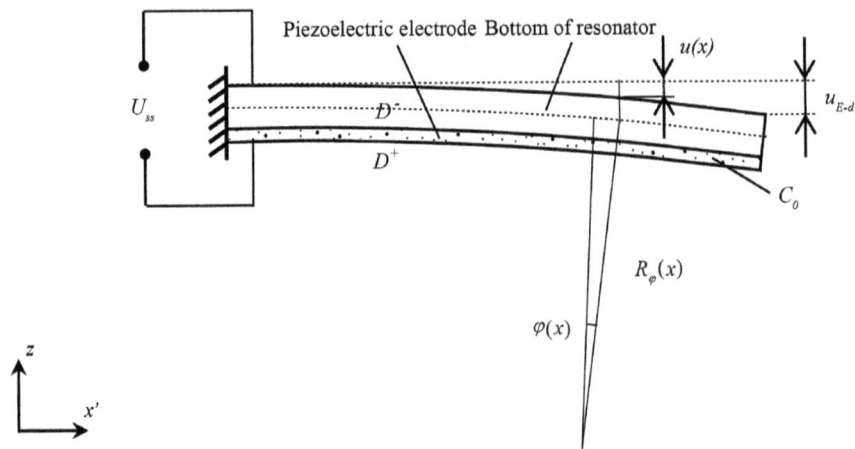

**Fig. 4.14**  Strain and signal of flexural piezoelectric sensor

$$S(x, t) = u_{E\_s} \frac{h_b + h_p}{(R - R_0)^2} \sin\left(\omega_p t - \beta_d + \frac{\pi}{2} - \beta_s\right) \quad (4.70)$$

Equation (4.70) is then substituted into Eq. (2.20) from Chap. 2, evaluating the detection voltage $U_s$ of the flexural piezoelectric sensor electrode, as shown below:

$$\begin{cases} U_{ss} = \frac{h_p e_{31} \int_0^{l_p} S_1(x) dx}{\varepsilon_{33} l_p} = \frac{e_{31} h_p \int_0^{R-R_0} u_{E\_d}(h_b+h_p)\sin(\omega_p t - \beta_d + \frac{\pi}{2} - \beta_s) dx}{\varepsilon_{33}(R-R_0)^3} \\ = \frac{u_{E\_s} e_{31} h_p (h_b+h_p)}{\varepsilon_{33}(R-R_0)^2} \sin\left(\omega_p t - \beta_d + \frac{\pi}{2} - \beta_s\right) \end{cases} \quad (4.71)$$

Equations (4.10), (4.58), and (4.64) are then substituted into Eq. (4.71), deriving the following statements:

$$U_{ss} = \frac{M_{c0}}{k_s^*} \frac{1}{\sqrt{(1 - v_s^2)^2 + 4\xi^2 v_s^2}} \frac{R - R_0}{2H} \frac{e_{31} h_p (h_b + h_p)}{\varepsilon_{33}(R - R_0)^2}$$

$$\sin\left(\omega_p t - \beta_d + \frac{\pi}{2} - \beta_s\right)$$

$$= \frac{\Omega U_{p0}\omega_p}{k_d^* k_s^* \sqrt{(1 - v_d^2)^2 + 4\xi^2 v_d^2}\sqrt{(1 - v_s^2)^2 + 4\xi^2 v_s^2}} \cdot \frac{e_{31}\rho}{\varepsilon_{33}}\left(\frac{c_{13}e_{33}}{c_{33}} - e_{31}\right)$$

$$\frac{\pi t_2 H R b_p h_p (R - R_0)(h_b + h_p)^2}{12(R - R_0)^2} \sin\left(\omega_p t - \beta_d + \frac{\pi}{2} - \beta_s\right)$$

$$= \Omega \frac{U_{p0}\omega_p}{k_d^* k_s^* \sqrt{(1 - v_d^2)^2 + 4\xi^2 v_d^2}\sqrt{(1 - v_s^2)^2 + 4\xi^2 v_s^2}}$$

$$K_G K_P \sin\left(\omega_p t - \beta_d + \frac{\pi}{2} - \beta_s\right) \quad (4.72)$$

where $K_G$ represents the influence coefficient of the geometric parameter and $K_P$ represents the influence coefficient of the physical parameter. Their expressions are as follows:

$$\begin{cases} K_G = \frac{\pi t H R b_p h_p (h_b+h_p)^2}{12(R-R_0)} \\ K_P = \frac{e_{31}\rho}{\varepsilon_{33}}\left(\frac{c_{13}e_{33}}{c_{33}} - e_{31}\right) \end{cases} \quad (4.73)$$

According to Eq. (4.72), under ideal conditions, the signal output from the flexural piezoelectric sensor, i.e. the gyroscope's output angular velocity signal $U_s$, has the same frequency as the input signal $U_d$ of the gyroscope's driving force, and their phase difference is $\pi/2$.

## 4.5  Sensitivity Analysis of CVGs

Sensitivity is critical to CVG performance and has significant influence on the gyroscope's scale factor. This factor has a great deal to do with the gyroscope's resolution and ultimately determines the micro-gyroscope's bias stability. Therefore, when a gyroscope is designed, a sense model must be established to analyze the influence of various sensitivity parameters [3].

### 4.5.1  Quality Factor Model

The quality factor is a key parameter for measuring a gyroscope's performance. It determines the gyroscope's detection sensitivity and response speed, while the magnitude of the quality factor is determined by the resonator's energy loss mechanism. In energy terms, the quality factor refers to the ratio of the total energy stored in the system to the energy lost during every period of oscillation. The less energy lost in each period, the higher the quality factor. Therefore, in a gyroscope, the higher its quality factor, the better it is.

When a CVG is working, the energy loss of its resonator determines the magnitude of its quality factor. Considering the definition and physical significance of the quality factor, the following can be derived:

$$Q = \frac{2\pi E}{\Delta E} = \frac{2\pi E}{\sum_{i=1}^{n} \Delta E_i} \tag{4.74}$$

where $E$ represents the total energy stored in the vibratory system, and $\Delta E_i$ represents the energy loss caused in the $i$th energy loss mode. Energy loss is an internal factor that affects the resonator's quality factor. Generally, the total energy loss of the resonator is divided into six types of basic energy loss: air damping loss $1/Q_{gas}$, surface defect loss $1/Q_{sur}$, support loss $1/Q_{sup}$, thermoelasticity loss $1/Q_{ther}$, internal friction loss $1/Q_{fri}$ and other environmental losses $1/Q_{other}$. The relationship between the total energy loss $1/Q$ and the above six sub-types of energy loss are shown below:

$$\frac{1}{Q} = \frac{1}{Q_{gas}} + \frac{1}{Q_{sur}} + \frac{1}{Q_{sup}} + \frac{1}{Q_{ther}} + \frac{1}{Q_{fri}} + \frac{1}{Q_{other}} \tag{4.75}$$

### I.  Impact of Surface Loss on the Quality Factor

In general, resonators are machined by turning or grinding techniques. This physical machining process inevitably damages the surface morphology of the shell, causing surface microcracks and burns. The crystalline grains on the resonator surface are distributed in random directions, so the thermal conductivity of the resonator surface is inhomogeneous, and inhomogeneous thermal conduction causes surface loss. The resonator surface loss is directly proportional to the thickness of the damaged layer.

T. Uchiyama et al. established the following CVG surface loss formula [4]:

$$\frac{1}{Q_{sur}} = 2h_{dam}\left(\frac{1}{L+l} + \frac{1}{R}\right)\frac{E\Gamma\gamma^2}{c}\frac{\omega_0\tau}{1+(\omega_0\tau)^2} \tag{4.76}$$

$\tau = \psi^2\frac{c}{\kappa}$, where $c$ represents the thermal capacity per unit volume; $\kappa$ represents the thermal conductivity; $\gamma$ represents the thermal expansion coefficient; $h_{dam}$ represents the thickness of the damaged layer; and $\Gamma$ represents the resonator temperature.

## II.  Impact of Thermoelasticity Loss on the Quality Factor

During the resonator's vibrations, the length of its middle plane does not change, but on the left and right sides of its middle plane, different temperature gradients are generated because the material is compressed and stretched. This internal temperature conduction causes a loss in thermoelastic damping. Thermoelastic damping in MEMS gyroscopes has received a high degree of attention from the industry, and most thermoelastic damping models are based on Zener's classic thermoelastic damping theory. Thermoelastic damping loss is closely related to the structural size of the resonator and is usually calculated according to another definition of the quality factor:

$$\frac{1}{Q_{ther}} = 2\left|\frac{\text{Im}(\omega_r)}{\text{Re}(\omega_r)}\right| \tag{4.77}$$

where $\omega_r$ represents the natural frequency of the resonant ring during thermoelastic vibrations, expressed in the following [5]:

$$\omega_r^2 = \frac{1}{\rho h_s}\left\{[D_s + F_1(\omega_r)][(n/R)^2 + \lambda_m^2]^2 + \frac{F_0(\omega_r)\lambda_m^2}{R(1+\varepsilon_r)} + \frac{K(1-\mu^2)\lambda_m^4}{R^2(1+\varepsilon_r)[(n/R)^2 + \lambda_m^2]^2}\right\} \tag{4.78}$$

where $\lambda_m = \lambda_L$, $\varepsilon_r$, $F_0(\omega_r)$ and $F_1(\omega_r)$ are formulas related to parameters such as the thermal expansion coefficient, specific heat per unit volume, and temperature. Their specific expressions can be found in the associated literature. Considering that the resonator sidewall is composed of a thin ring and a thick ring, its total thermoelastic damping loss can be separately calculated.

The thermoelastic damping of the resonator base plate is [6]:

$$\frac{1}{Q_{thb}} = \Delta_D\left[\frac{6}{\xi^2} - \frac{6}{\xi^3}\left(\frac{\sinh\xi + \sin\xi}{\cosh\xi + \cos\xi}\right)\right] \tag{4.79}$$

where $\xi = h\sqrt{\frac{\omega_3\rho c}{2\kappa}}$, $\omega_3 = \frac{q_n}{R^2}\sqrt{\frac{D_b}{\rho h_b}}$, $q_n$ is a frequency constant. They can be found in a handbook of mechanical vibrations: $D_b = \frac{Eh_b^3}{12(1-\mu^2)}$, $\Delta_D = \frac{(1+\mu)\gamma\beta T}{\rho c}$, $\beta = \frac{E\gamma}{1-2\mu}$.

Therefore, the resonator's total thermoelastic damping loss is:

$$\frac{1}{Q_{ther}} = \frac{\frac{U_s}{Q_{rs}} + \frac{U_r}{Q_{rr}} + \frac{U_b}{Q_{rb}}}{U_s + U_r + U_b} \tag{4.80}$$

### III.   Impact of Support Loss on the Quality Factor

The support loss is part of the vibrational energy, i.e., the energy transmitted through the support structure to the base. As shown in Fig. 4.15, the elastic base plate of the resonator is connected to its base, and its vibratory action on the base can be modeled as a discrete acting force and moment. If the energy flow transmitted to the base isn't reflected, it can be expressed as:

$$\Pi = \frac{1}{2}\text{Re}(F \cdot V) \tag{4.81}$$

where $F$ represents the vectorial load, and $V$ represents the resonant angular velocity at that point.

The energy loss caused by the support loss can be calculated according to the following equation:

$$U_{sup} = \Pi \frac{2\pi}{\omega} \tag{4.82}$$

The base is treated as a thick plate with a thickness of $h_p$; shear stress of $F_z$; flexural moment of $M_b$, and torsional moment of $M_{b\theta}$ perpendicular to the plate are considered. They can be evaluated based on the theory of plates and shells:

$$M_b(r) = -D_b\left[\frac{\partial^2 u_b}{\partial r^2} + \mu\left(\frac{1}{r}\frac{\partial u_b}{\partial r} + \frac{1}{r^2}\frac{\partial^2 w}{\partial \theta^2}\right)\right]$$

$$M_{b\theta}(r) = -D_b\left[\left(\frac{1}{r}\frac{\partial u_b}{\partial r} + \frac{1}{r^2}\frac{\partial^2 w}{\partial \theta^2}\right) + \mu\frac{\partial^2 u_b}{\partial r^2}\right]$$

$$F_x(r) = -D_b\frac{\partial}{\partial r}\nabla^2 u_b \tag{4.83}$$

where $\nabla^2 = \frac{\partial^2}{\partial r^2} + \frac{1}{r}\frac{\partial}{\partial r} + \frac{\partial^2}{r^2\partial\theta^2}$.

**Fig. 4.15** Support loss model of the resonator

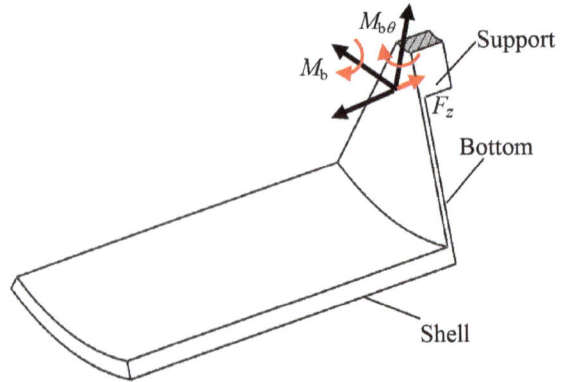

$r = r_0$ is then substituted into Eq. (4.83) to obtain the load on the base. The relationship of the load to the normal angular velocity $\Omega_b$, tangential angular velocity $\Omega_{b\theta}$, and linear velocity $V_z$ can be revealed from the equation below:

$$
\begin{bmatrix} \Omega_b \\ \Omega_{b\theta} \\ V_z \end{bmatrix} = \frac{1}{\sqrt{\rho h_b D_p}} \begin{bmatrix} y_{11}k^2 & 0 & 0 \\ 0 & y_{22}k^2 & y_{23}k \\ 0 & y_{32}k & y_{33} \end{bmatrix} \begin{bmatrix} \overline{M}_b \\ \overline{M}_{b\theta} \\ \overline{F}_z \end{bmatrix} \tag{4.84}
$$

where $k$ represents the number of free waves at frequency $\omega$, $k = \left[\omega(\rho h_p/D_p)^{1/2}\right]^{1/2}$, $D_p = \frac{Eh_p^3}{12(1-\mu^2)}$; $\overline{F}_z$, $\overline{M}_b$, and $\overline{M}_{b\theta}$ are the integral of shear stress $F_z$, flexural moment $M_b$, and torsional moment $M_{b\theta}$ respectively. $\overline{F}_z = \int_0^{\pi/4} F_z d\theta$, $\overline{M}_b = \int_0^{\pi/4} M_b d\theta$, $\overline{M}_{b\theta} = \int_0^{\pi/4} M_{b\theta} d\theta$. $y_{11}, y_{22}, y_{23}, y_{32}$, and $y_{33}$ are already calculated in the literature [7]. Therefore, the energy flow transmitted to the base can be evaluated:

$$
\Pi_{Fz} = \frac{2}{\sqrt{\rho h_p D_p}}(y_{32}k\overline{M}_{b\theta}\overline{F}_z + y_{33}\overline{F}_z^2)
$$

$$
\Pi_{Mb\theta} = \frac{2}{\sqrt{\rho h_p D_p}}(y_{22}k^2\overline{M}_{b\theta}^2 + y_{23}k\overline{F}_z\overline{M}_{b\theta})
$$

$$
\Pi_{Mb} = \frac{2}{\sqrt{\rho h_p D_p}}y_{11}k^2\overline{M}_b^2 \tag{4.85}
$$

The support damping loss caused to the resonator is finally evaluated according to the definition of damping loss:

$$
\frac{1}{Q_{sup}} = \frac{U_{sup}}{2\pi S} = \frac{\Pi_{Fz} + \Pi_{Mb} + \Pi_{Mb\theta}}{\omega S} \tag{4.86}
$$

IV.   **Impact of Internal Friction Loss on the Quality Factor**

In addition to elastic deformation, the resonator material also has some inelastic deformation. While vibrating, the resonator has to overcome viscosity by doing useless work, which causes a loss of internal friction in the resonator. In the Kelvin-Voigt model, material damping $\xi$ should be expressed as follows according to Hooke's law:

$$
\sigma = E\left(\varepsilon + \xi\frac{d\varepsilon}{dt}\right) \tag{4.87}
$$

where $\sigma$ and $\varepsilon$ represent the stress and strain of the elastic infinitesimal body in the non-middle plane. It is evident that $\zeta d\varepsilon/dt$ causes a loss in viscous internal friction, and its corresponding moment acts on the cross-section of the infinitesimal body, as shown in Fig. 4.16.

The deformation of the infinitesimal body $\varepsilon$ is:

**Fig. 4.16** Internal friction
loss model

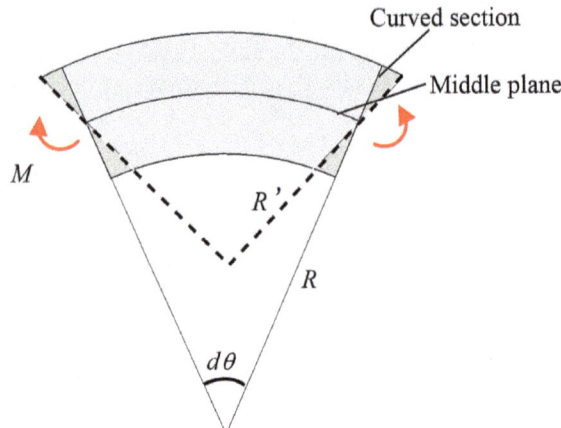

$$\varepsilon = \frac{(R' + l_r)(d\theta + \delta d\theta) - (R + l_r)d\theta}{(R + l_r)d\theta} \approx \frac{l_r \delta d\theta}{R d\theta} \tag{4.88}$$

where $R'$ represents the new radius of the non-middle plane after deformation; and $l_r$ represents the distance from the middle plane. Note that the length of the middle plane is constant during vibrations, therefore:

$$Rd\theta = R'(d\theta + \delta d\theta) \tag{4.89}$$

And let $\Delta k = \frac{\delta d\theta}{R d\theta}$, so,

$$\varepsilon = \Delta k l_r \tag{4.90}$$

where $\Delta k$ represents the change of the curvature:

$$\Delta k = \frac{1}{R'} - \frac{1}{R} = \frac{\partial \varphi}{R d\theta} \tag{4.91}$$

where $\varphi$ represents the angular deflection of the infinitesimal body.

For the force generated by the viscous damping effect, $\sigma' = E\xi\dot{\varepsilon}$, moment $M$ is generated in the element interface, as depicted in the following equation:

$$M = \int \sigma' l ds$$
$$= \frac{EI\xi}{R} \frac{\partial \dot{\varphi}}{\partial \theta} = \frac{K\xi}{R} \frac{\partial \dot{\varphi}}{\partial \theta} \tag{4.92}$$

where $I$ represents the moment of inertia of the cross section, and its magnitude is determined by section thickness $h$:

$$I = \frac{h^3}{12}dx \tag{4.93}$$

The work done by moment $M$ does not contribute to the vibrations of the resonator, so it is useless work. In the side wall of the resonator, its dissipation energy $U_{sr}$ is divided into two parts, which exist on the resonant ring and support ring, and these are respectively written as:

$$U_{sr} = \frac{1}{2} \int_0^l \int_0^{2\pi} \frac{h_s^3}{12} \frac{E\xi}{R^3} (\dot{v}' - \dot{w}'')(v' - w'')d\theta dx$$

$$+ \frac{1}{2} \int_l^{L+l} \int_0^{2\pi} \frac{h_r^3}{12} \frac{E\xi}{R^3} (\dot{v}' - \dot{w}'')(v' - w'')d\theta dx \tag{4.94}$$

It is noted that the resonator not only has radial deformation but also axial deformation. In the axial deformation, the infinitesimal inertia is expressed as:

$$I = \frac{h^3}{12}Rd\theta \tag{4.95}$$

Accordingly, the change of the axial deformation curvature is:

$$\Delta k_x = \frac{W_1(x)''}{(1 + W_1(x)'^2)^{3/2}} \cos(\omega t) \tag{4.96}$$

Therefore, the dissipation energy caused by the resonator's axial deformation is:

$$U_{sx} = \frac{1}{2} \int_0^l \int_0^{2\pi} \frac{h_s^3}{12} E\xi \Delta \dot{k}_x \Delta k_x Rd\theta dx + \frac{1}{2} \int_l^{L+l} \int_0^{2\pi} \frac{h_r^3}{12} E\xi \Delta \dot{k}_x \Delta k_x Rd\theta dx \tag{4.97}$$

The internal friction loss caused at the bottom of the resonator is then considered. The infinitesimal inertia at the bottom is:

$$I_b = \frac{h_b^3}{12}rd\theta \tag{4.98}$$

The corresponding change of the bottom curvature is:

$$\Delta k_b = \frac{u_b(x)''}{(1 + u_b(x)'^2)^{3/2}} \cos(\omega t) \tag{4.99}$$

Therefore, the dissipation energy at the bottom of the resonator is:

$$U_{sb} = \frac{1}{2} \int_{r_0}^{R} \int_{0}^{2\pi} \frac{h_b^3}{12} E\xi \Delta \dot{k}_b \Delta k_b r d\theta dr \qquad (4.100)$$

According to the established definitions, the loss caused by the internal friction of the material is ultimately evaluated as follows

$$\frac{1}{Q_{fri}} = \frac{U_{sx} + U_{sr} + U_{sb}}{2\pi S} \qquad (4.101)$$

### 4.5.2   Angular Velocity Sense Model

Under normal circumstances, when a CVG is working in its drive mode, it is driven by a self-excited oscillation circuit. According to the principle of self-excited oscillations, the frequency of the exciting voltage of is identical to the frequency of the resonator in drive mode, so $\omega_p = \omega_d$. When angular velocity is input into the gyroscope, the detection voltage $U_s$ output from the piezoelectric electrode is:

$$U_s = \Omega \frac{U_{p0}\omega_p}{k_d^* k_s^* \sqrt{(1 - v_d^2)^2 + 4\xi^2 v_d^2} \sqrt{(1 - v_s^2)^2 + 4\xi^2 v_s^2}}$$
$$K_G K_P \sin\left(\omega_p t - \beta_d + \frac{\pi}{2} - \beta_s\right)$$
$$= \Omega \frac{U_{p0}\omega_d}{k_d^* k_s^* 2\xi \sqrt{(1 - v_s^2)^2 + 4\xi^2 v_s^2}} K_G K_P \sin\left(\omega_p t - \beta_s\right) \qquad (4.102)$$

In the gyroscope's measurement and control circuit, the detection voltage output from the piezoelectric electrode must be demodulated by the drive signal $U_{p0}\sin \omega_p t$ to acquire the angular velocity signals, therefore:

$$U_\Omega = U_s \cdot U_d = \Omega \frac{U_{p0}^2 \omega_d}{k_d^* k_s^* 2\xi \sqrt{(1 - v_s^2)^2 + 4\xi^2 v_s^2}}$$
$$K_G K_P \sin\left(\omega_p t - \beta_s\right) \sin \omega_p t$$
$$= \Omega \frac{U_{p0}^2 \omega_d}{k_d^* k_s^* \xi \sqrt{(1 - v_s^2)^2 + 4\xi^2 v_s^2}} K_G K_P \left[\cos\left(2\omega_p t - \beta_s\right) - \cos \beta_s\right] \qquad (4.103)$$

The amplitude of the angular velocity signals output from the gyroscope can be obtained by filtering high-frequency signals from the above signals:

$$U_\Omega = \Omega \frac{U_{p0}^2 \omega_d}{k_d^* k_s^* \xi \sqrt{(1 - v_s^2)^2 + 4\xi^2 v_s^2}} K_G K_P |\cos \beta_s| \tag{4.104}$$

The sensitivity of a CVG is equal to the ratio of the amplitude of the angular velocity signals output from the gyroscope to the amplitude of the gyroscope's input angular velocity. According to Eq. (4.104), the angular velocity sensitivity $S_g$ of the CVG is:

$$S_g = \frac{U_\Omega}{\Omega} = \frac{U_{p0}^2 \omega_d}{k_d^* k_s^* \xi \sqrt{(1 - v_s^2)^2 + 4\xi^2 v_s^2}} K_G K_P |\cos \beta_s| \tag{4.105}$$

In a second-order system, its quality factor has the following relationship with its damping ratio:

$$Q = \frac{1}{2\xi} \tag{4.106}$$

Under normal conditions, since a CVG has excellent structural symmetry, the difference in stiffness between its drive mode and sense mode can be ignored, i.e. $k_d^* = k_s^*$, therefore Eq. (4.105) can be rewritten as:

$$S_g = \frac{U_\Omega}{\Omega} = \frac{2Q^2 U_{p0}^2}{\sqrt{k_d^{*3} m_d^*} \sqrt{Q^2(1 - v_s^2)^2 + v_s^2}} K_G K_P |\cos \beta_s| \tag{4.107}$$

Similarly, after a 90° signal phase shift, the detection voltage output from the piezoelectric electrode for detection generates signal $U_{p0} \cos \omega_p t$, and by demodulating this signal, we can obtain an angular velocity signal. This allows for the angular velocity sensitivity of the gyroscope to ultimately be obtained:

$$S_g = \frac{U_\Omega}{\Omega} = \frac{2Q^2 U_{p0}^2}{\sqrt{k_d^{*3} m_d^*} \sqrt{Q^2(1 - v_s^2)^2 + v_s^2}} K_G K_P |\sin \beta_s| \tag{4.108}$$

From Eqs. (4.107) and (4.108), the angular velocity sensitivity of the CVG can be shown to be related to the resonator's geometric parameters, physical parameters, and driving voltage magnitude and frequency, as well as the mode of the angular velocity signal demodulation.

The specific expressions of the parameters, including $k_d^*$, $m_d^*$, $K_G$, and $K_P$, are substituted into Eq. (4.108), with the following equation derived:

$$S_g = \frac{Q^2}{\sqrt{Q^2(1 - v_s^2)^2 + v_s^2}} 12 \sqrt{\frac{\pi \rho}{E^3}} \left(\frac{8}{4 - \pi}\right)^{3/2} \frac{e_{31}}{\varepsilon_{33}} \left(\frac{c_{13} e_{33}}{c_{33}} - e_{31}\right)$$

$$\frac{R^5 b_p h_p \left(h_b + h_p\right)^2}{(R - R_0)^4 \left(t H^3\right)^{3/2}}$$

$$\cdot \frac{t H^5}{\left[5 H^3 t + 3 H t (R - R_0)^2\right]^{1/2}} |\cos \beta_s| \qquad (4.109)$$

From Eq. (4.109), it can be seen that the CVG's angular velocity sensitivity increases as the density of the metal resonator material decreases, that it decreases as the elasticity modulus $E$ of the metal material increases, and that it increases as the piezoelectric strain coefficient $e_{31}$ ($e_{31}$ is normally a negative value) and as $e_{33}$ increases.

After the structural size and material type of the resonator are determined, if the resonator's geometric and physical parameters have exact effects on sensitivity, the main factors that determine the sensitivity of the gyroscope are the amplitude $U_{d0}$ of the driving voltage and the quality factor $Q$ of the resonator. The gyroscope's sensitivity is directly proportional to the amplitude $U_{d0}$ of the driving voltage. According to the self-excited driving principle of CVGs, the gyroscope's driving voltage frequency is equal to the frequency of the drive mode. Another major factor that affects gyroscope sensitivity is the difference between the resonator's drive mode frequency and its sense mode frequency, i.e., the resonator frequency split $\Delta \omega = \omega_d - \omega_s$, or the ratio of the drive mode frequency to the sense mode frequency $\upsilon_s = \omega_p / \omega_s$.

Taking a resonator with a working mode frequency of 4000 Hz as an example, the different frequency split $\Delta \omega$ and quality factor $Q$ of the resonator can be obtained from as based on Eqs. (4.108) and (4.109), and are shown to have a relationship with the gyroscope's sensitivity, as shown in Fig. 4.17.

As shown in Fig. 4.17, if the drive signal is directly demodulated, the angular velocity sensitivity of the gyroscope increases as the $Q$ value of the resonator increases, and there is an irregular relationship between the gyroscope's frequency

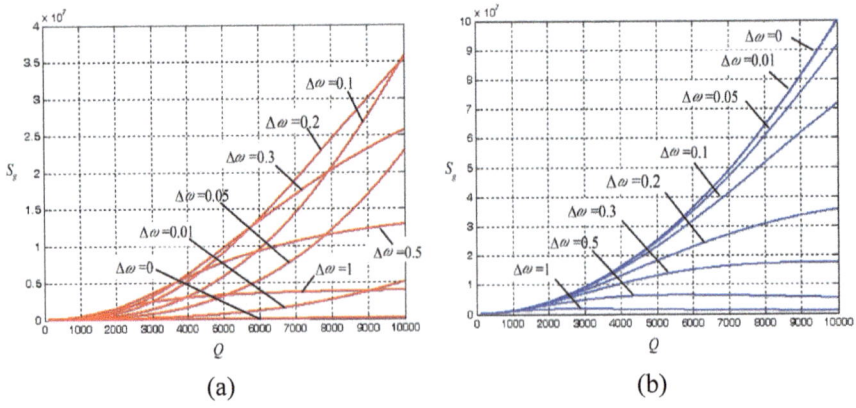

**Fig. 4.17** Relationship between the quality factor of the resonator and the gyroscope's sensitivity. **a** Drive signal modulation; **b** drive signal modulation after a 90° phase shift

split and its sensitivity. If a 90° phase-shift demodulation is adopted for processing a CVG's drive signals, the angular velocity sensitivity of the CVG increases as its frequency split decreases. When the frequency split is below 0.5 Hz, the angular velocity sensitivity increases as the $Q$ value increases; when the frequency split is below 0.01 Hz, the gyroscope sensitivity becomes saturated. Relatively speaking, a higher absolute value can be obtained for gyroscope sensitivity through the 90° phase-shift demodulation method as compared to the direct signal demodulation method. Therefore, to obtain higher angular velocity sensitivity, the drive mode frequency and sense mode frequency of the CVG resonator must be very close to each other (frequency split of less than 0.1 Hz). The resonator should have a high mechanical quality factor ($Q$ value greater than 5000), and the angular velocity signal of the gyroscope should be demodulated after a 90° drive signal phase shift.

# References

1. Tao, Y. (2011). *Research on the key technologies of cup-shaped wave gyros*. National University of Defense Technology.
2. Sun, S. (1997). *Theoretical mechanics*. National University of Defense Technology Press.
3. Xi, X. (2014). *Research on the drift mechanisms and suppression technologies of cup-shaped wave gyroscopes*. National University of Defense Technology Press.
4. Uchiyama, T., Tomaru, T., Tobar, M., Tatsumi, D., Miyoki, S., Ohashi, M., Kuroda, K., Suzuki, T., Sato, N., & Haruyama, T. (1999). Mechanical quality factor of a cryogenic sapphire test mass for gravitational wave detectors. *Physics Letters A, 261*(1), 5–11.
5. Lu, P., Lee, H., Lu, C., & Chen, H. (2008). Thermoelastic damping in cylindrical shells with application to tubular oscillator structures. *International Journal of Mechanical Sciences, 50*(3), 501–512.
6. Sun, Y., & Tohmyoh, H. (2009). Thermoelastic damping of the axisymmetric vibration of circular plate resonators. *Journal of Sound and Vibration, 319*(1), 392–405.
7. Kauffmann, C. (1998). Input mobilities and power flows for edge-excited, semi-infinite plates. *The Journal of the Acoustical Society of America, 103*(4), 1874–1884.

# Chapter 5
# Manufacturing of CVGs

The high performance of CVGs means that their resonators have high standards for manufacturing precision. This chapter introduces the basic process of manufacturing metal resonator structures, which are based on the material characteristics and structural dimensions of the resonator.

## 5.1  Resonator Materials

### 5.1.1  Material Characteristics

The primary functional characteristics of high-performance shell vibratory gyroscopes are their high sensitivity and wide operating temperature range. This high performance requires that the resonator have high manufacturing precision, a high mechanical quality factor, and a low frequency-temperature coefficient. Some natural materials, such as silicon, quartz, and sapphires, have extremely high mechanical quality factors, extremely low frequency temperature coefficients, and stable physicochemical properties. At present, hemispherical resonator gyroscopes are the only shell vibratory gyroscope to possess standard inertial accuracy. Their resonators are made of fused quartz, an isotropic material that has the same extremely high mechanical quality factor and extremely low frequency-temperature coefficient as quartz crystals, completely ensuring gyroscope performance. However, since fused quartz is a hard and brittle material that is very difficult to machine, it is characterized by its high requirements for precision machining equipment, high machining costs, and low machining efficiency.

To reduce machining costs and improve machining efficiency, metal can be used to make low-cost CVG resonators. When an alloy is selected as the resonator material, a great deal of attention should be paid to the alloy's internal friction, linear thermal expansion coefficient, and the temperature coefficient of the elastic modulus.

© National Defense Industry Press 2021

X. Wu et al., *Cylindrical Vibratory Gyroscope*, Springer Tracts in Mechanical Engineering, https://doi.org/10.1007/978-981-16-2726-2_5

An elastic alloy is a type of alloy with special elastic properties, excellent mechanical performance, and some special physicochemical properties. It is often used for the production of various elastic components applied in instruments, automatic devices, and precision machines. The properties of elastic alloys are divided into three types: elastic properties, non-elastic properties, and elastic anomalies. Generally speaking, the elastic modulus $E$ and $G$ of metal decrease as temperature increases. The corresponding temperature coefficients of the elastic modulus are $\beta_E$ and $\beta_G$. The characteristic of a constant modulus alloy is that its $E$ or $G$ values, or the resonant frequency $f_0$ of the manufactured alloy, does not change the temperature outside of a certain range. The phenomenon in which the elastic modulus increases with temperature is known as elastic anomaly. If the elastic anomaly can make up for the decrease of the rise in the normal elastic modulus temperature within a certain temperature range, constant elasticity can be obtained in this range in order to produce an alloy with a constant modulus. Taking the iron-nickel-based constant modulus alloy 3J58 as an example, its mass fraction of nickel is 43.0–43.6%, which provides the alloy with constant elasticity. The elastic modulus of a constant elastic alloy has an essentially constant value within normal temperature ranges. It is generally specified that the temperature coefficient of the elastic modulus is $\beta_E < 120 \times 10^{-6}/°C$ from −60 to 100 °C. Common alloys with constant elasticity include Fe–Ni-Cr and Fe–Ni-Mo ferromagnetic alloys, Mn-Cu anti-ferromagnetic alloys, and Nb-Zr paramagnetic alloys [1].

Constant elastic alloys are ideal materials for manufacturing resonators. This type of alloy has a low temperature coefficient for its elastic modulus, a low frequency-temperature coefficient and thermal expansion coefficient, high elasticity and strength, a low elastic aftereffect, and high corrosion resistance. It is particularly worth noting that, given their high plasticity, such alloys are easy to machine into various elastic components with complex structures. According to various possible applications, the material composition and thermal processing techniques of these alloys can be adjusted to meet the requirements of the elastic modulus' temperature coefficient and frequency-temperature coefficient. Common high-performance intermetallic compound-strengthened constant modulus alloys include Chinese-made 3J53, 3J58, and 3J59, as well as the US-made alloy Ni-Span-C 902. Their respective constituent materials and properties are shown in Tables 5.1 and 5.2.

**Table 5.1** The main constituents of typical constant elastic modulus alloys

| Alloy | Chemical constituents (%) | | | | | | | |
|---|---|---|---|---|---|---|---|---|
| | Ni | Cr | Ti | Al | Mo | Mn | C | Fe |
| 3J53 | 42 | 5.5 | 2.5 | 0.75 | | <0.7 | <0.05 | Remainder |
| 3J58 | 43 | 5.5 | 2.5 | 0.6 | | <0.7 | <0.05 | Remainder |
| 3J59 | 43.5 | 5.0 | 2.5 | 0.5 | 0.5 | <0.5 | <0.03 | Remainder |
| Ni-Span-C alloy 902 | 41–43.5 | 4.9–5.7 | 2.2–2.7 | 0.3–0.8 | | <0.8 | <0.06 | Remainder |

**Table 5.2** Main properties of typical constant elastic modulus alloys

| 合金<br>Alloy | Operating temperature range (/°C) | Linear thermal expansion coefficient ($10^{-6}$/°C) | Mechanical quality factor | Frequency-temperature coefficient ($10^{-6}$/°C) | Main features |
|---|---|---|---|---|---|
| 3J53 | −40 to 80 | 8.3 | ≥10,000 | 0 to 20 | Wide operating range, low thermal expansion coefficient. And low frequency-temperature coefficient. The disadvantage is that it is sensitive to changes in its constituent materials |
| 3J58 | −40 to 120 | 8.3 | ≥10,000 | −5 to 5 | |
| 3J59 | −40 to 120 | 8.3 | ≥18,000 | −2 to 2 | |
| Ni-Span-C alloy 902 | −45 to 70 | 7.6 | ≥20,000 | −5 to 5 | |

The temperature coefficient of the elastic modulus of constant modulus alloys is very sensitive to changes in its material constituents, especially toward nickel content. The relationship between the nickel content and the elastic module temperature coefficient is shown in the hand book [2]. The elastic module temperature coefficient of constant modulus alloys is also related to the alloy's operating frequency. The mechanical quality factor of constant modulus alloys is also sensitive to material constituents, and this type of constant modulus alloy has a controllable $Q$ value. The degree of aluminum and silicon content also has a major influence on the alloy's mechanical quality factor.

## 5.1.2 Material Processing

Before an alloy is used, it generally needs to be cold-processed to significantly reduce its surface area, leading to differences in textures between the material surface and the center (Fig. 5.1). The differences are ultimately manifested through the heterogeneity of the material's physical properties, which greatly reduce the material's mechanical quality [3]. A reasonable selection of thermal treatment and cold processing techniques is vital to improving the performance of the resonator material, because the thermal treatment of a constant modulus alloy does not merely strengthen the alloy or fix the shape of the elastic component. Instead, it adjusts the distribution of various elements between the base metal and the metallic compounds, and controls the alloy structure so as to fully stabilize the resonator's frequency-temperature coefficient and mechanical quality factor.

Common thermal treatment technologies for constant modulus alloys [4]:

(1) Solid solution treatment: The solid solution temperature of an alloy is generally between 950 and 980 °C. After 15–40 min of being maintained at this constant temperature, the alloy is quickly quenched in water to produce a

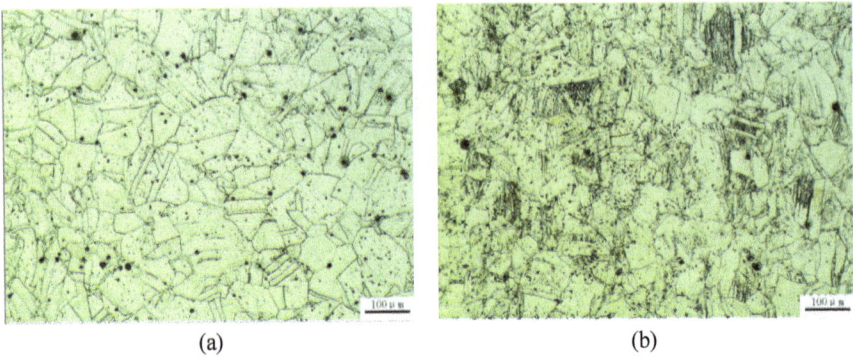

(a)                                                          (b)

**Fig. 5.1** Metallographic diagram of various parts of a constant modulus alloy 3J53. **a** Metallographic diagram of bar core; **b** metallographic diagram of bar edge

single-phase austenite structure with micro-crystalline grains. If the temperature is too low, the solid solution will be incomplete; if the temperature is too high, the crystalline grains will be long, large, and uneven. As a result, the alloy's machinability and its performance after its aging treatment will both be adversely affected. In order to ensure a complete solid solution, the machining temperature should be kept as low as possible.

(2)    Aging treatment: The aging treatment should be performed after the solid solution or cold deformation processing in order to reduce the nickel content in the alloy matrix, with a resulting change in ferromagnetism. This is an important condition for ensuring both high mechanical properties and constant elastic properties. In the aging treatment process, appropriate thermal treatment parameters are selected and the precipitation quantity and distributional pattern of the dispersed phases are controlled to produce excellent properties in the desired areas. After the alloy is cold-rolled, lattice distortion will be caused by material deformation. After undergoing aging treatment, the lattice distortion inside the material is restored. At this time, the disperse $\gamma'$ phase is precipitated and embedded in crystal lattices, obstructing dislocation migration, reducing plastic deformation, and improving material hardness and elasticity. Observation of the dispersion phase precipitation in a material's metallographic structure helps to confirm the properties of the material, thus revealing the influence of the thermal treatment on the material's material properties.

Within the aging temperature range, the strengthening effect increases proportionally with the aging temperature. After undergoing solid solution and aging treatment, the alloy is best strengthened at about 700 °C. The temperature and frequency coefficients of the elastic modulus increases with the aging temperature, shifting from a negative value to a positive value, reaches its maximum from 650 to 700 °C, and then returns to a negative value after the aging temperature exceeds 700 °C.

(3) Aging: Whether a constant modulus alloy is applied statically or dynamically, it must undergo aging after its aging treatment in order to ensure high time stability and low inelastic behavior among the parts in use. If aging is not performed, the resonant frequency of the mechanical filter resonator will increase over time.

(4) Cold deformation: Cold working deformation promotes the precipitation and homogenization of the strengthening phase, so after cold deformation aging is performed, the alloy acquires better mechanical properties and constant elastic properties. The higher the rate of cold deformation, the more obvious the strengthening effect is. Moreover, after aging strengthening is performed, the maximum peak temperature generally ranges from 50 to 100 °C lower than the temperature of the alloy as a solid solution. The cold deformation rate of alloy strips is generally 30–70%, while that of the wires is 30–75%.

The elastic module temperature coefficient of a cold-deformed alloy is a negative value, which gradually turns into a positive value as the aging temperature rises, but the coefficient returns to a negative value after the aging temperature exceeds 700 °C. Alloys used as frequency elements should undergo appropriate cold deformation before aging treatment. Moreover, the aging temperature must be strictly controlled to obtain a low frequency-temperature coefficient and a high mechanical quality factor.

(5) Tempering: Tempering is performed in a protective atmosphere, and its goal is to keep the element surface bright and clean. The alloy is tempered after quenching or quenching and cold deformation at above 500 °C. Due to the dispersion effect, a dispersed phase is precipitated inside the alloy, while the nickel content in the matrix is reduced, thus improving the strength and elasticity of the alloy and turning the temperature coefficient of the elastic modulus from a negative value into a positive value and makes it approach zero. At this time, the ferromagnetism of the alloy matrix is also changed.

(6) Thermal treatment in a protective atmosphere and vacuum thermal treatment: During the thermal treatment of metal, a protective atmosphere is generally used, or a vacuum thermal treatment is performed instead. This is because metal materials are in a chemically active state and are easily oxidized at high temperatures. Thermal treatment in a protective atmosphere or vacuum-based thermal treatment can be adopted to effectively prevent metallic materials from being oxidized and can promote the decomposition of oxides on the alloy surface, thus improving the performance of alloys after thermal treatment. Commonly used protective gases include nitrogen, hydrogen, argon, and water vapor (Table 5.3).

Taking the thermal treatment of alloy 3J58 as an example, 3J58 is a type of Fe–Ni–Cr–Ti precipitate-strengthened austenitic stainless steel alloy, and both its elastic module temperature coefficient and frequency-temperature coefficient are large even before thermal treatment is applied. After the solution heat treatment of an alloy (>950 °C), a $\gamma'$-phase saturated solid solution will be obtained, and its lattice type is turned into a face-centered cubic type. The alloy is then thermally treated at a low

**Table 5.3**  Mechanical properties of 3J53 and 3J58 alloys in different states

| Alloy | State | E/Gpa | G/Gpa | Poisson's ratio | Q |
|-------|-------|-------|-------|-----------------|---|
| 3J53 | Solid solution + aging | 176–186 | 61–70 | – | 10,000 |
|  | Cold deformation + aging | 181–206 | 67–74 | 0.30–0.45 | |
| 3J58 | Cold deformation + aging | 181–206 | 67–74 | 0.30–0.45 | 10,000 |

temperature to decompose the solid solution within it in order to precipitate different types of dispersed phases along the grain boundary and matrix, enabling the matrix to be significantly strengthened and the alloy to be greatly strengthened, and to obtain a high $Q$ value and a low elastic lag. At this time, the temperature coefficient of the elastic modulus is adjusted to a value close to zero so that the material performance approaches constant elasticity. If the alloy is cold-deformed (>40%) prior to its aging treatment, which will promote the precipitation and distribution of its dispersed phases, and its properties will be greatly improved.

## 5.2  Resonator Manufacturing Techniques

### 5.2.1  Basic Technical Procedures

If the ratio of the wall thickness of a component to the inner curvature radius (or overall dimensions) is less than 1:20, it is referred to as a thin-walled component. The CVG resonator is a typical gyratory thin-walled structure (shell wall thickness is about 0.5–1 mm). Deformation is a prominent problem arising from the processing of thin-walled components, and this deformation is primarily divided into the following three types. First, force deformation. Under the action of the clamping force, components deform very easily, owing to the low thickness of the workpiece's wall. As a result, both the size and shape accuracy of the workpiece are affected. Second is thermal deformation. Also, because of its thin wall, the workpiece deforms as the cutting heat increases. As a result, the size of the workpiece is difficult to control. Third, vibration deformation. Under the action of the cutting force, especially the radial cutting force, a thin-walled component is very inclined to vibrate and deform. As a result, not only are the dimensional accuracy and shape of the workpiece affected, but its positional precision and surface roughness are as well. Poor control over these factors tends to cause machining problems such as deformation, oversized pieces, and poor surface quality, and these seriously affect the component's machining precision and finished product ratio.

The precision machining of a resonator's metal structure mainly involves the machining of the inner and outer circles of the bottom of the resonator. Due to the relatively low stiffness of its thin-walled structure, the resonator is vulnerable to machining deformation under the impact of the cutting force, cutting heat, and cutting vibrations. This machining deformation has significant impact on the machining

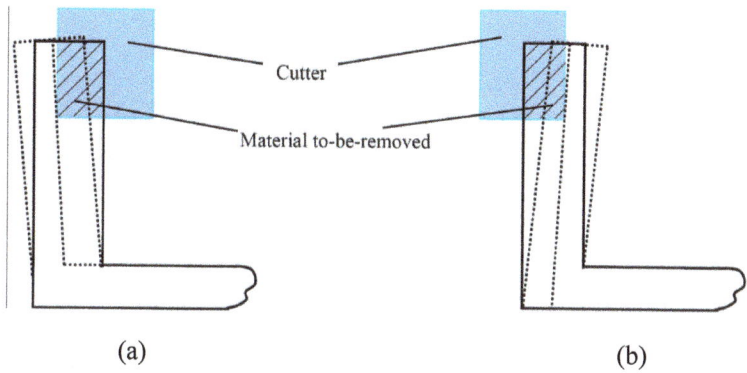

**Fig. 5.2** Machining deformation of the resonator shell wall. **a** Deformation of inner circle; **b** deformation of outer circle

precision and efficiency of the resonator. Figure 5.2 shows the principle of machine deformation in the resonator's metal structure. The main cause of machining errors is "cutter back-off" caused by the thin wall becoming deformed during machining. These errors often make the upper part of the wall thicker than the lower part of the wall. Therefore, during the precision machining of the resonator's metal structure, the issue of machining deformation must be solved, and relevant machining parameters, such as the cutting speed, feed rate, tool geometric parameters, fixture location mode, and clamp force must all be determined.

Figure 5.3 shows the resonator's rough metal structure and finished structure. For every to-be-finished surface of the resonator's metal structure, there is an allowance of 0.6–1 mm for finishing to be performed on the blank to ensure the effect of thermal treatment on the rough piece and to give it high stiffness to avoid machining deformation during the early stages of precision machining.

Precision machining can be performed through the following steps [5]:

(1) Finish the sleeve of the inner circle. The nominal size of the outer circle of the sleeve should be the same as the nominal size of the inner circle of the resonator.

**Fig. 5.3** Blank and finished structure of the resonator

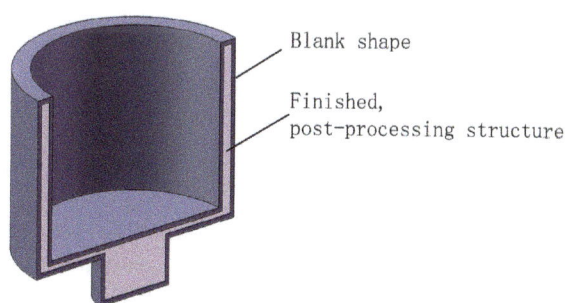

**Fig. 5.4** Finished metal
structure of the resonator

(2)  Finish the inner circle, inner bottom, and top surface of the resonator. The size
of the inner circle of the resonator should be subject to the size of the outer
circle of the sleeve of the inner circle, and a basic axis system should be adopted
for transitional fitting.

(3)  Clamp the sleeve of the inner circle to the inner bottom surface of the resonator.

(4)  Finish the outer circle, outer bottom, and support bar of the resonator. After
turning, cut off the support bar and remove the component.

Figure 5.4 shows the finished metal structure of the resonator.

According to the mechanical quality factor characteristics of constant modulus
alloys, the surface state of an alloy material can be enhanced to improve the material's
mechanical quality factor. Therefore, when the metal structure of the resonator is
machined, not only should the shape accuracy and positional precision be ensured,
the high machining quality of the component surface must also be assured. After
cutting is complete, the resonator can be polished with a flexible polishing tool and
abrasive particles or with the use of another polishing medium to effectively improve
the resonator's surface quality and release the residual stress on the resonator surface
to improve the stability of the resonator.

### 5.2.2  Nickel-Based Alloy Machining Methods

A metal resonator is often made of iron-nickel-based constant modulus alloy, which is
a ferromagnetic precipitate-strengthened constant modulus alloy. As for the mechanical properties of iron-nickel alloys, it has high plasticity in solid solution states and
can be made into a complex-shaped elastic element through cold stamping. After

proper aging treatment, the alloy will acquire high strength, a high $Q$ value, and low elastic lag, so alloys of this type are widely used. After aging treatment is performed following solid solution treatment or cold straining, it will be strengthened and gain excellent constant elastic properties. This type of alloy has excellent properties, including a low temperature coefficient (or frequency-temperature coefficient) for its elastic modulus, a high mechanical quality factor, a high degree of wave velocity conformity, high strength and elastic modulus, low elastic aftereffect and lag, low linear expansiveness, and high corrosion resistance. Owing to the high nickel content (43.0%-43.6%) in iron-nickel-based constant modulus alloys and the alloys' high working plasticity, machining problems such as severe adhesive wear are unavoidable, but can be improved by the following methods:

(1)  Control the impact of work hardening. The high degree of work hardening is the most distinct feature of nickel base alloy machining, and is the primary problem that needs to be resolved in the cutting process. We must be fully aware of the harm caused by work hardening and try to avoid and reduce its impact. One approach is to increase the back cutting depth and feed rate above the work hardening layer, and to try to prevent the cutting edge from cutting the work hardening layer and keep the feed state unchanged. Another method is to perform the aging treatment of workpieces after rough machining in order to lower its plasticity to reduce the impact of work hardening.

(2)  Select an appropriate amount for cutting. Increasing the back cutting depth or feed rate is bound to expand the cutting area, thus enhancing the deformation force and frictional force, which also enhances the cutting force. Maintaining a low cutting speed, a high, constant feed rate, and a moderate back cutting depth is useful for maintaining continuous feed. It's best to try to prevent the cutting edge from cutting the work hardening layer so as to minimize the impact of work hardening.

(3)  Select a reasonable geometric angle for the cutter. Because the rake face bears the cutting force, the rake angle has the greatest impact on the cutting force. Increasing the rake angle can reduce cutting deformation and cutting force, preventing the generation of cutting heat and lowering the cutting temperature, but a decrease in the heat conduction area of the cutter head is bound to increase the cutting temperature. The main cutting force changes by about 1.5% for each 1° change in the rake angle. The impact of the rake angle on the cutting force decreases with as the cutting speed increases. The flank face affects the service life of the cutter and the effect of work hardening. After rapid wear, the flank face is certain to induce serious work hardening; the major function of the relief angle is to reduce the friction between the flank face and the machining surface. Therefore, serious work hardening will be caused if the relief angle is excessively small and the cutting force is too large. The cutter will wear more quickly if the relief angle is excessively large and the cutter strength is low.

(4)  Select a proper cutting fluid. A cutting fluid can be used in the cutting process to reduce the cutting force. The work consumed in the cutting process is mainly used to overcome the deformation of the metal and the cutting friction between

the cutter and machined material. When used correctly, the cutting fluid can reduce the friction during this process. Taking steel material machining as an example, the cutting friction caused on the rake face consumes about 35% of the work; the friction between the workpiece and the flank face consumes about 5–15% of the work. If the cutter is sufficiently cooled with a cutting fluid, the cutting force can be reduced by more than 30%. Regardless of whether rough turning or fine turning is used, a sufficient cooling effect must be ensured in order to make machining proceed smoothly. A great deal of cutting heat will be generated during rough machining if there is a great deal of cutting performed. At this time, the cutting temperature must be reduced, so a cutting fluid with excellent cooling performance should be selected, such as an ionic-type cutting fluid or 3–5% emulsion. During precision machining, the major function of the cutting fluid is to reduce the workpiece's surface roughness and to improve machining precision, so extreme pressure cutting oil or 10–12% extreme pressure emulsion can be selected for this. In addition, the reasonable addition of an additive to reduce the cutting fluid's surface tension reduction can enable the cutting fluid to penetrate into the metal micro-cracks in the plastically deforming area in order to lower the strengthening coefficient and reduce the cutting force, thereby making the cutting process easier.

(5)  Select an appropriate cutter material. Common cutter materials include cemented carbides, ceramics, and cubic boron nitride (CBN). A common cemented carbide has a hardness of 89–93 HRA, as well as a high abrasive resistance and heat resistance. However, as a result of its poor bending strength and fracture toughness, cemented carbide cannot withstand strenuous cutting vibrations or high impact loads.

There are two common ceramic tool materials: $Al_2O_3$-based ceramic and $Si_3N_4$-based ceramic. A cutter made of $Al_2O_3$-based ceramic has a hardness of 91–95 HRA, higher than that of cemented carbides. When used with care, it is highly durable. It also has high heat resistance and chemical stability, and a low friction coefficient, adhesion, and surface roughness. However, like a cemented carbide cutter, the greatest fault of cutters made of $Al_2O_3$-based ceramic is that they have low bending strength and poor impact resilience.

CBN is converted from hexagonal boron nitrides (HBN) treated with a catalyst at high temperature and pressure. This is a new type of cutter material that came into use in 1970s. CBN has a high hardness and abrasive resistance. Its micro-hardness can reach 8000–9000 HV, close to the hardness of diamond. With its high thermal stability (up to 1400 °C) and chemical inertness, it undergoes hardly any chemical reactions at high temperatures. With a low surface friction coefficient, a cutter made of CBN can significantly reduce the degree of adhesion in the cutting process. Therefore, CBN is an ideal cutter material for cutting resonators made of an iron-nickel-based constant modulus alloy.

## 5.2.3 Resonator Cutting

The machinability of a workpiece material depends mainly on a material's mechanical and physical properties (e.g., hardness, strength, plasticity, toughness, and thermal conductivity). Moreover, the chemical constituents, metallographic structure, and micro-hardness of the material also have some influence on the material's machinability.

During the precision machining of the resonator's metal structure, the precision, production cost, and efficiency of the machining process must be ensured. Critically, processing pathways must be suitable for mass production. The CVG resonator is a gyratory thin-walled component and is made of an iron-nickel-based constant modulus alloy. Resonator cutting is related to its geometric accuracy and the material's physical properties. Due to their thin walls, thin-walled components easily deform during machining. Due to their high nickel content and material plasticity, iron-nickel-based constant modulus alloys are vulnerable to severe cementation during machining, and large cutting forces are generated. Moreover, the cutter becomes seriously worn as a result, which affects the machining precision. The cutting process can be optimized by the following methods:

(1) Select a proper cutter material and geometric parameters. For thin-walled workpiece finishing, the shank hardness must be high and the cutting edge must be sharp. A finishing cutter with a chisel edge can be used. The rake face, flank face, and circular point profile of the finishing cutter must be ground with oilstone. During the cutting process, the first force to deform the component is the radial cutting force. The magnitude of the radial cutting force borne by the component in the cutting process is directly related to the cutter used and to the amount of turning, and a cutter with a large cutting edge angle for its tool should be adopted whenever possible. The magnitude of the rake angle determines the degree of cutting deformation and the sharpness of the rake angle. If the rake angle is large, cutting deformation and friction will be low. If the cutting force is small but the rake angle is too large, the wedge angle of the cutter is small, the cutter strength is low, the heat dissipation performance of the cutter is poor and the wear rate is high. Increasing the rake angle of the cutter can reduce the frictional force, and the cutting force will be reduced as well. However, an excessively large rake angle also weakens the cutter strength, so various factors should be considered when a cutter is chosen.

(2) Select a proper amount for cutting. The cutting amount also has a crucial impact on the magnitude of the cutting force. The magnitude of the cutting force is closely related to the cutting amount. When the back cutting depth and feed rate increase together, the cutting force will increase as well, causing a large amount of deformation. This is extremely unfavorable to the turning of thin-walled components. Although the cutting force will decrease somewhat when the back cutting depth is reduced while the feed rate is raised, the residual area on the workpiece surface increases, and so does the surface roughness. This increases the inner strain of low-strength thin-walled components, thus

deforming the component. Therefore, a larger amount of cutting can be selected for rough machining in order to complete the machining as quickly as possible and minimize inter-process cutting. This is because the cutting stress generated during rough machining can be completely eliminated by thermally treating the component. In component finishing, a small cutting depth and low feed rate must be selected. In addition, a cutter with correct geometric parameters should be used so as to minimize the impact of the cutting force on the component. If the cutting power is unchanged, the cutting force can also be reduced by increasing the cutting speed, but after the cutting speed increases, massive frictional heat will accumulate on the lower cutting layer, raising the cutting temperature, so the cutting speed cannot be greatly raised.

In addition, high-speed machining or ultraacoustic vibration cutting techniques can be used to effectively suppress the impact of dynamic cutting vibrations on the machining quality of thin-walled components. Compared with conventional cutting, high-speed cutting force is reduced by at least 30%. The radial cutting force is especially greatly reduced. In addition, during the high-speed cutting process, more than 95% of cutting heat is removed from the workpiece by cutting action, accumulating less heat on the workpiece, so the workpiece is subjected to warping or expansion deformation at high temperature.

## 5.2.4  Machining Error Analysis

The machining deformation of thin-walled components is caused by many factors, including cutting parameters and cutting methods, cutter properties (e.g., material, stiffness, and geometric parameters), machine tool properties (e.g., stiffness and machining precision), cooling conditions, cutting vibrations, and other random factors, all of which have an impact on the rate of deformation and the surface quality. The main factors include the cutting force, cutting heat, material, clamping, and residual stress.

### I.  Influence of Clamping

Clamping is a method by which a machine tool is linked to thin-walled components. No matter which machining method is adopted, the clamping of thin-walled components during machining is the primary condition under which thin-walled component machining is performed. 20% to 60% of machining errors are caused by improper clamping. Specifically, the clamping procedure, clamp point location, and clamping force may cause different degrees of deformation in thin-walled components as well as different machining errors.

Due to high requirements for final machining precision, the allowance given to precision machining is quite small, and rough machining errors will be reflected in the finished workpieces through error reflection. In addition, during the machining process, the clamping force and cutting force fluctuate, producing a coupling effect, which leads to a redistribution of the machining-induced residual stress and initial residual stress in the component, worsening the deformation of the workpiece. Therefore, the improvement of the clamping process is very important to controlling the machining deformation of thin-walled components.

The workpiece is easily deformed under the clamping force, and its dimensional accuracy and shape accuracy are affected as a result. Figure 5.5a shows a three-edged circle on the inner surface of a thin-walled component cut on a three-jaw chuck. As shown in Fig. 5.5b, the workpiece deforms under the clamping force of the three-jaw chuck; i.e., there is elastic deformation at the point of contact between the workpiece and the chuck. A circular inner hole obtained from cutting is shown in Fig. 5.5c. If the size of the hole is measured while the chuck is not loosened, it meets the dimensional requirements specified in the drawing, but because hole turning is earlier than elastic deformation, after the chuck is loosened, the outer circle of the workpiece will be restored while the inner hole will be become three-edged. Figure 5.6 shows the measurement results of the roundness of the inner surface of an early-machined resonator, and in it there is an obvious three-edged circle error.

Therefore, when a thin-walled workpiece is turned, radial clamping must be avoided as much as possible, and axial clamping is preferred. When the workpiece is clamped on the end face of axial clamping sleeve, the direction of the clamping force is changed so that the clamping force is distributed along the axial direction of the workpiece, but the workpiece has high axial stiffness which prevents the clamping force from generating clamping deformation. In addition, when performing precision machining of thin-walled components made of optical crystals, a vacuum chuck or pneumatic chuck is usually used for low-stress clamping in order to effectively reduce deformation.

## II.  Impact of the Cutting Force on Machining

The generation of the cutting force is attributed to chip deformation. Moreover, the cutting force directly affects the generation of cutting heat, thus affecting the abrasive resistance and durability of cutters as well as their chip curling, chip breaking, and surface machining quality. Cutting heat is one of the most important physical phenomena in the cutting process. In terms of the energy consumed in the cutting

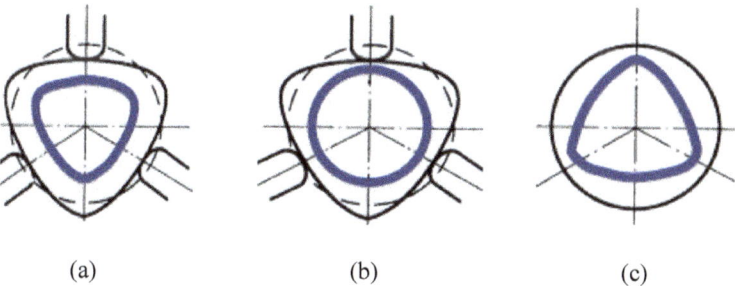

(a)  (b)  (c)

**Fig. 5.5** Impact of the cutting clamping force. **a** Pre-machining; **b** inner hole cutting; **c** post-machining

**Fig. 5.6** Three-edged
roundness error on the inner
surface of the resonator

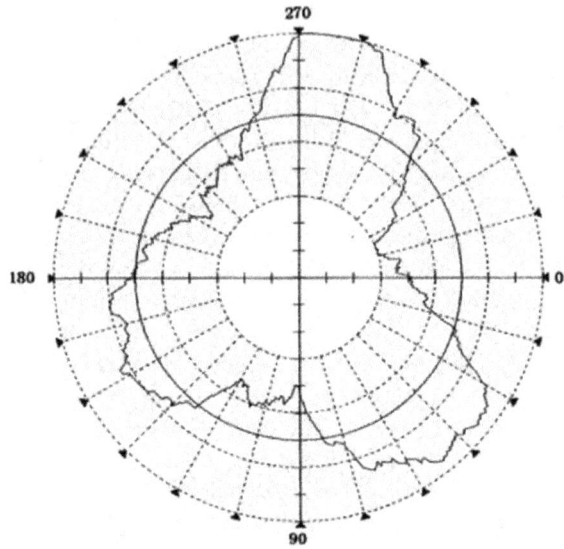

process, 98–99% is converted into thermal energy, while the remaining 1–2% is used to form a new surface or latent energy in the form of lattice distortion and other forms. Massive cutting heat raises the cutting temperature, which directly affects the machining quality. The heat in the cutting area is dissipated through chips, workpieces, cutters, and the surrounding media. Thin-walled components are easily deformed by thermal expansion during machining, due to their low wall thickness and high thermal expansion coefficient, and this deformation has a significant impact on the amount of cutting, resulting in a difference between the post-cooling finished size and the designed size.

Considering the low stiffness of thin-walled components, the cutting force can cause rebounding deformation in the component. If the cutting force is very large, greater than the elastic limit of the material, the workpiece will be subjected to extrusion deformation; i.e., plastic deformation. Besides, during the cutting process, the thin-walled component chatters under the cutting force, and the fluctuant cutting force also appears on the machined surface of the workpiece in the form of rippling, affecting the surface quality and thus weakening the workpiece performance. Moreover, because of a difference in stiffness among different parts of thin-walled components, error reflection occurs in the cutting process, making it difficult to guarantee the component's machining precision. The workpiece deforms in the direction of the cutting force action, leading to a change in the relative position of the workpiece and the cutter, as well as in the amount of cutting and the back cutting depth, thus affecting the shape accuracy of the workpiece.

### III.    **Impact of Residual Stress**

Residual stress refers to a type of stress that maintains balance inside a component in the absence of any load. Residual stress is composed of two parts: initial residual

**Fig. 5.7** Extruding
deformation of workpiece
surface caused in the cutting
process

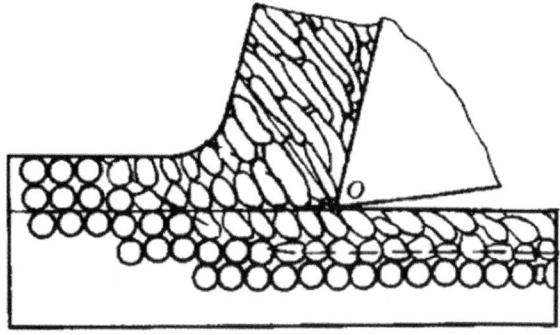

stress and machining-induced residual stress produced under the action of the cutting force and cutting heat. Because the rear surface of the cutter compresses the machined surface, a third deformation zone comes into being during the cutting process, and this is shown in Fig. 5.7. As a result of the deformation, the crystalline grains on the workpiece surface are stretched or even fibrillated, and the volume of the surface layer changes as well, producing a residual stress on the surface of the workpiece. In general, the residual stress on the surface of the workpiece cannot be averted. In addition, the residual stress on the surface of the workpiece is gradually released over time, causing a distribution of this stress across the surface material of the workpiece, thus deforming the workpiece.

The initial residual stress of a thin-walled component is related to the material of the blank, while the machining-induced residual stress is generated during the machining of the thin-walled component and easily deforms the component.

The residual stress on the resonator surface affects the gyroscope's performance in two aspects: first, the homogeneity of the physical properties of the resonator material; second, the resonator shape, which gradually changes with the release of the residual stress. The greater the residual stress on the surface of the resonator, the more seriously the surface layer is deformed than the inner layer. That is, the higher the inhomogeneity of the physical properties of the surface layer and inner layer, the lower the vibration stability of the resonator. As the residual stress is released, the shape and size of the resonator change, thereby destabilizing its performance, which gradually deteriorates over time.

An X-ray diffraction stress meter can be used to measure the stress. When blades with different cutter angles are adopted, each type of blade is given a unique feed rate. We can then observe how the cutter angles influence the formation of residual stresses on the resonator surface. Table 5.4 shows the residual stresses on the workpiece surface measured under different cutting conditions.

To prevent the surface residual stress from affecting the resonator's bias stability, the surface deformation of the workpiece should be minimized; i.e., the residual stress on the workpiece surface should be reduced as much as possible. The methods for eliminating the residual stress on the rough workpiece surface include natural aging,

**Table 5.4** Measurement results of residual stresses under different cutting conditions

| Feed rate<br>Blade model | F70 | F60 | F50 | F40 | F30 |
|---|---|---|---|---|---|
| CCGT0602-SC | −223.24 | −242.91 | −271.61 | −308.38 | −352.57 |
| CCGT0602-FX | 91.48 | −34.96 | −88.89 | −13.46 | −86.55 |
| DCGT0702-SC | 11.53 | −34.25 | −89.18 | −166.32 | −232.33 |
| DCGT0702-FX | 19.55 | 15.88 | 13.37 | −52.81 | −165.96 |
| DCMT0702-SU | 79.38 | 42.55 | −22.34 | −51.54 | −76.90 |

Units of residual stress: MPa

vibration aging, hammering aging (hammering method), thermal aging treatment, and ultraacoustic peening.

## 5.3   Trimming of Imperfect Resonators

### 5.3.1   Frequency Split of Imperfect Resonators

The manufacturing errors of the resonator mainly include roundness errors, which occur on the inner and outer circles of the resonator shell wall, along with coaxiality errors. These errors cause uneven distribution in the quality and stiffness distribution of the vibration structure. A research study by Kerimov et al. shows that the technological defects of a wave gyroscope resonator cause the formation of two natural axis systems in the resonator at an angle of 45° to each other [6, 7]. While vibrating these two axes, the resonator has different natural frequencies, and the frequency difference is called the natural frequency split. As shown in Fig. 5.8, non-ideal CVG

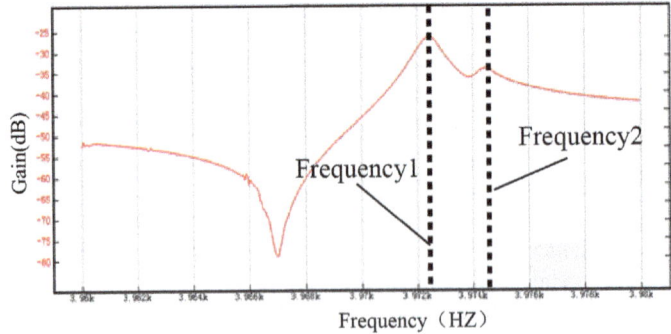

**Fig. 5.8** Frequency response diagram of a resonator with a frequency split

**Fig. 5.9** Resonator model
with imperfect mass points

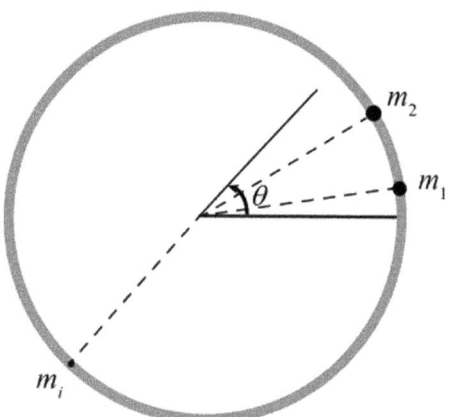

resonators always have a frequency split in their working modes, while the value of
the frequency split varies between different technological defect types and levels.

A ring model is used to build a theoretical model for the frequency split of
resonators. Considering that a continuous mass change can always be discretized into
multiple mass points, an imperfect mass point model is used analyze the frequency
split of the resonator. As shown in Fig. 5.9, when mass points are distributed
arbitrarily, the kinetic variation of the resonator is:

$$\Delta T_s = \sum_{i=1}^{n} \frac{1}{2} m_i \left( \dot{u}_i^2 + \dot{v}_i^2 + \dot{w}_i^2 \right) \tag{5.1}$$

where $m_i$ represents the $i$th mass point; $u_i$, $v_i$, and $w_i$ respectively represent the axial,
tangential, and radial velocities at this point. Considering that the added point is
actually very small, the amplitude of the resonator is not affected, but a change takes
place in the mode shape direction. The vibration of the resonator is then re-expressed
as:

$$\begin{cases} u = U(x) \cos 2(\theta - \varphi) cos(\omega t) \\ v = V(x) \sin 2(\theta - \varphi) cos(\omega t) \\ w = W(x) \cos 2(\theta - \varphi) cos(\omega t) \end{cases} \tag{5.2}$$

where $\varphi$ represents the angle of deflection of the mode shape after mass addition.

Therefore, the newly formed natural frequency $\omega_{sp}$ is [8]:

$$\omega_{sp}^2$$

$$= \frac{2S}{\rho R h_r \int_l^{L+l} \int_0^{2\pi} [U(x)^2 + W(x)^2]\cos^2 2\theta\,dx\,d\theta + \rho R h_s \int_0^l \int_0^{2\pi} [U(x)^2 + W(x)^2]\cos^2 2\theta\,dx\,d\theta}$$
$$+ \sum_{i=1}^{n} m_i[U(X_i)^2 + W(X_i)^2]\cos^2 2(\theta_i - \varphi)$$

$$= \omega_0^2 \left\{ 1 + \frac{\sum_{i=1}^{n} \frac{1}{2}m_i[U(X_i)^2 + W(X_i)^2]\cos^2 2(\theta_i - \varphi)}{\rho R h_r \int_l^{L+l} \int_0^{2\pi} [U(x)^2 + W(x)^2]\cos^2 2\theta\,dx\,d\theta + \rho R h_s \int_0^l \int_0^{2\pi} [U(x)^2 + W(x)^2]\cos^2 2\theta\,dx\,d\theta} \right\}^{-1} \tag{5.3}$$

where $(X_i, \theta_i)$ represents the coordinate position of point mass.

In Eq. (5.3), if the magnitude and direction of the added mass are known quantities, $\varphi$ will be the only uncertain quantity, so:

$$\frac{\partial \omega_{sp}^2}{\partial \varphi} = 0 \tag{5.4}$$

Equation (5.3) is then substituted into Eq. (5.4):

$$\tan 2n\varphi = \frac{\sum_{i=1}^{n} m_i[U(X_i)^2 + W(X_i)^2]\sin 2n\theta_i}{\sum_{i=1}^{n} m_i[U(X_i)^2 + W(X_i)^2]\cos 2n\theta_i} \tag{5.5}$$

If the distribution of mass points is highly consistent, the above equation can be simplified as follows:

$$\tan 2n\varphi = \frac{\sum_{i=1}^{n} m_i \sin 2n\theta_i}{\sum_{i=1}^{n} m_i \cos 2n\theta_i} \tag{5.6}$$

The following are the two solutions of the parameter $\varphi$ in Eq. (5.6):

$$\varphi_1 = \frac{\arctan\left\{ \dfrac{\sum_{i=1}^{n} m_i[U(X_i)^2 + W(X_i)^2]\sin 2n\theta_i}{\sum_{i=1}^{n} m_i[U(X_i)^2 + W(X_i)^2]\cos 2n\theta_i} \right\}}{2n}$$

$$\varphi_2 = \frac{\arctan\left\{ \dfrac{\sum_{i=1}^{n} m_i[U(X_i)^2 + W(X_i)^2]\sin 2n\theta_i}{\sum_{i=1}^{n} m_i[U(X_i)^2 + W(X_i)^2]\cos 2n\theta_i} \right\}}{2n} + \frac{\pi}{2n} \tag{5.7}$$

In specific terms, it shows that uneven mass distribution results in two mode shapes that are at an angle of 45° to each other. The higher frequency value $\omega_H$ and the lower frequency value $\omega_L$ can be separately expressed as:

$$\omega_H^2 = \omega_0^2 \left\{ 1 + \frac{\sum\limits_{i=1}^{n} \frac{1}{2} m_i [U(X_i)^2 + W(X_i)^2] \cos^2 2(\theta_i - \varphi_1)}{\rho R h_r \int_l^{L+l} \int_0^{2\pi} [U(x)^2 + W(x)^2] \cos^2 2\theta dx d\theta + \rho R h_s \int_0^l \int_0^{2\pi} [U(x)^2 + W(x)^2] \cos^2 2\theta dx d\theta} \right\}^{-1}$$

$$\omega_L^2 = \omega_0^2 \left\{ 1 + \frac{\sum\limits_{i=1}^{n} \frac{1}{2} m_i [U(X_i)^2 + W(X_i)^2] \sin^2 2(\theta_i - \varphi_1)}{\rho R h_r \int_l^{L+l} \int_0^{2\pi} [U(x)^2 + W(x)^2] \cos^2 2\theta dx d\theta + \rho R h_s \int_0^l \int_0^{2\pi} [U(x)^2 + W(x)^2] \cos^2 2\theta dx d\theta} \right\}^{-1} \quad (5.8)$$

Therefore, the newly formed frequency is codetermined by the magnitude and direction of the mass. Moreover, a maximum value and a minimum value are formed in two directions at an angle of 45° to each other. This means that a frequency split occurs.

It is not difficult to see that the fourth-harmonic component generated by the uneven distribution of the resonator mass is the main part of the resonator's frequency split. If the material density and uneven elastic modulus distribution are not considered, a geometric error model can be built for the resonator's shell wall based on the error shape of the fourth harmonic, as shown in Fig. 5.10. The amplitude of the fourth harmonic is equal to the roundness error of the inner and outer circles, while the eccentric error between the center of the outer circle and the center of the inner circle is known as the coaxiality error.

The finite element software ANSYS is used to build an error-based finite element (FE) model for resonators, as shown in Fig. 5.12. A modal analysis incorporating machining errors is conducted on the resonator model to study the impact of machining errors on the resonator's dynamic characteristics. The geometric and physical parameters of the resonator are set as follows: $R = 12.5$ mm, $\rho = 8050$ kg/m$^3$, $E = 210$ MPa, $\mu = 0.3$.

A model is established for the resonator shell wall. The model contains inner and outer roundness errors and coaxiality errors as well as ideal geometric dimensions. After FE modal simulation, the relationship between the machining errors of the resonator and its frequency split can be expressed and are shown in Table 5.5.

The existence of the frequency split makes the drive mode and sense mode of the CVG resonator differ in frequency, exciting other resonator modes except for the principal modes, causing an orthogonal error and gyroscopic drift. Therefore, to

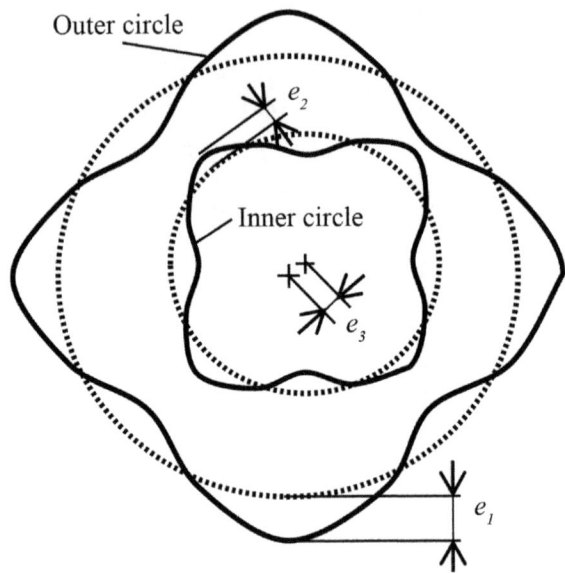

**Fig. 5.10** Inner and outer roundness error and coaxiality error of resonator

**Table 5.5** Results of the resonator machining error simulation and its frequency split

| Frequency split $\Delta f$ (Hz) | | Roundness error $e_1$, $e_2$ ($\mu$m) | | | |
|---|---|---|---|---|---|
| | | 1 | 3 | 5 | 10 |
| $e_3$ ($\mu$m) Coaxiality error $e_3$ ($\mu$m) | 5 | 0.171 | 0.472 | 1.214 | 2.063 |
| | 10 | 0.284 | 0.884 | 1.877 | 3.225 |
| | 15 | 0.638 | 1.224 | 2.452 | 5.471 |

create a high-precision CVG, machining precision must be improved or mass balance must be achieved to reduce the resonator's frequency split.

## 5.3.2   Frequency Trimming of Imperfect Resonators

Errors in machining the resonator's metal structure may cause uneven distribution in the mass and stiffness of the resonator, destroying the axial symmetry of the resonator structure, thereby creating a frequency split. The main factors affecting gyroscope performance are orthogonal errors caused by modal deflection and frequency split, and these must be eliminated or suppressed in order to achieve dynamic balance within the gyroscope. In the production of high-performance shell vibratory gyroscopes, the resonator's frequency split must be controlled below 0.01 Hz. The most common and effective way to achieve gyroscope balance is to control the mechanical balance of the resonator. This technology can correct the modal deflection angle of the resonator and can eliminate frequency split. It is therefore a key technology for improving gyroscope performance.

The mechanical balance of the resonator is divided into static balance and dynamic balance. Static balance is the basis of the resonator's mechanical balance, and is aimed at making the mass center of the resonator coincident with its axis of rotation in order to eliminate mass eccentricity. Static balance usually generates an additional frequency split. Dynamic balance is the key to the resonator's mechanical balance, and is aimed at reducing the resonator's frequency split and modal deflection angle within a certain range in accuracy. The dynamic balance of the resonator appears after the static balance, and the dynamic balance process shouldn't destroy the static balance. Therefore, the trimming of the resonator's dynamic balance should maintain central symmetry at its point of origin. The traditional method for adjusting the gyroscope resonator's mechanical balance is used to change the resonator's inertial mass and stiffness through material removal by making a serrated groove in the resonant ring and trimming it to achieve static and dynamic balance in the resonator.

Considering that both mass and stiffness changes may affect the resonant frequency of the resonator, the frequency split of the resonator can be trimmed by adjusting its mass or stiffness.

## I.   Mass-Type Frequency Split Trimming

Mass-type frequency split trimming refers to changing the frequency split by removing the local mass of the resonator to achieve resonator balance. The mass is generally changed by removing material from the resonator.

According to the operability of actual trimming, the present section lists two major methods of material removal. One way is to make a hole in the resonator to move the mass. Another way is to make a groove in the resonator, as shown in Fig. 5.11. There is a difference between the two trimming methods: hole machining only allows the implementation of trimming near the middle plane, which the maintains structural

**Fig. 5.11** Different methods of resonator trimming

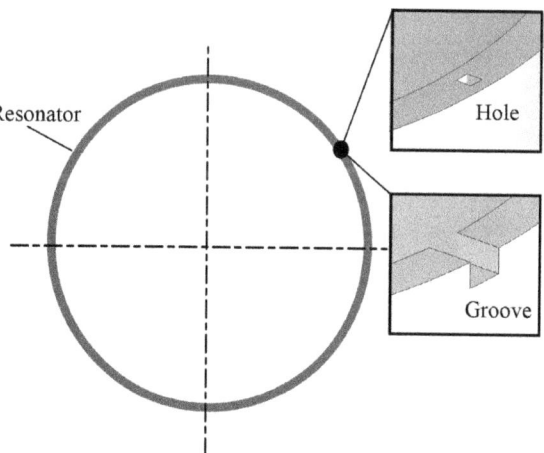

continuity of the resonator. However, groove machining damages the local structural continuity of the resonator.

FE modal simulation shows that a high-frequency axis is formed when hole machining is performed, while a low-frequency axis is formed where groove machining is performed. According to the fundamental formula of the natural frequency of the resonator:

$$\omega = \sqrt{\frac{K^*}{m^*}} \tag{5.9}$$

It can be found that a high-frequency axis is formed when the equivalent mass $m^*$ is reduced, while a low-frequency axis is formed under the condition that the equivalent

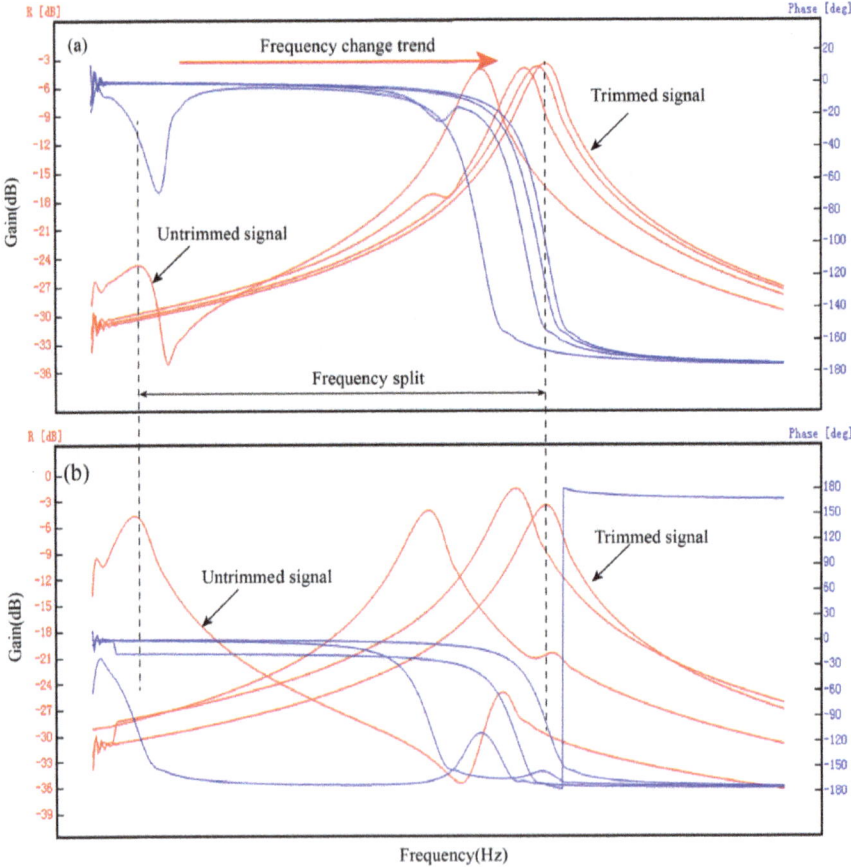

(a) Frequency response in drive mode  (b) Frequency response in sense mode

**Fig. 5.12**  Mass-type frequency split trimming

stiffness $K^*$ decreases. Therefore, it is inferred that hole machining realizes mass-type frequency split trimming, while groove machining realizes stiffness-type frequency split trimming.

According to the preceding section, the frequency split generated during mass unbalance is as follows:

$$
\omega_H^2 = \omega_0^2 \left\{ 1 + \frac{\sum_{i=1}^{n} \frac{1}{2} m_i [U(X_i)^2 + W(X_i)^2] \cos^2 2(\theta_i - \varphi_1)}{\rho R h_r \int_l^{L+l} \int_0^{2\pi} [U(x)^2 + W(x)^2] \cos^2 2\theta dx d\theta + \rho R h_s \int_0^l \int_0^{2\pi} [U(x)^2 + W(x)^2] \cos^2 2\theta dx d\theta} \right\}^{-1}
$$

$$
\omega_L^2 = \omega_0^2 \left\{ 1 + \frac{\sum_{i=1}^{n} \frac{1}{2} m_i [U(X_i)^2 + W(X_i)^2] \sin^2 2(\theta_i - \varphi_1)}{\rho R h_r \int_l^{L+l} \int_0^{2\pi} [U(x)^2 + W(x)^2] \cos^2 2\theta dx d\theta + \rho R h_s \int_0^l \int_0^{2\pi} [U(x)^2 + W(x)^2] \cos^2 2\theta dx d\theta} \right\}^{-1} \tag{5.10}
$$

where $(X_i, \theta_i)$ represents the coordinate position of the unbalance point mass.

The purpose of the mass balance is to make it tenable that $\omega_H = \omega_L$ by removing the point mass to a particular position, as follows:

$$
\sum_{i=1}^{n} \frac{1}{2} m_i [U(X_i)^2 + W(X_i)^2] \cos^2 2(\theta_i - \varphi_1)
$$

$$
- \sum_{j=1}^{n_2} \frac{1}{2} m_j [U(X_j)^2 + W(X_j)^2] \cos^2 2(\theta_j - \varphi_1)
$$

$$
= \sum_{i=1}^{n} \frac{1}{2} m_i [U(X_i)^2 + W(X_i)^2] \sin^2 2(\theta_i - \varphi_1)
$$

$$
- \sum_{j=1}^{n_2} \frac{1}{2} m_j [U(X_j)^2 + W(X_j)^2] \sin^2 2(\theta_j - \varphi_1) \tag{5.11}
$$

where $m_j$ represents the $j$th point of the mass removed; $U(x_j)$ and $W(x_j)$ represent the axial and radial amplitudes at the point. Equation (5.11) is valid because the frequency split can be eliminated by removing materials of different masses from a given position in the resonator. This illustrates the flexibility of the resonator's mass balance. If this mass is directly removed from the low-frequency axis (at $\theta = 0°, 90°$, 180° and 270°), magnitude $m$ at the removed point mass can be further simplified as:

$$m = \frac{\sum_{i=1}^{n} \frac{1}{2} m_i [U(X_i)^2 + W(X_i)^2] \cos^2 2(\theta_i - \varphi_1) - \sum_{i=1}^{n} \frac{1}{2} m_i [U(X_i)^2 + W(X_i)^2] \sin^2 2(\theta_i - \varphi_1)}{2(U_0^2 + W_0^2)} \quad (5.12)$$

A frequency trimming experiment can then be performed according to the above equation. This process and its results are shown in Fig. 5.12.

In the trimming experiment, the resonator's rigid-axis azimuth is identified first, and then a hole is made in the low-frequency axis for mass removal. Figure 5.12a, b respectively show the frequency response signal detected at the exciting electrode and detection electrode. It is evident that the modal vibrations of Imperfect Resonators can be detected at different frequencies, and the difference between these frequencies is called the frequency split. The resonator has an initial frequency split of 4.95 Hz, and after mass removal is performed, the natural frequency of the resonator's low-frequency axis constantly rises until it equalizes with the natural frequency of the high-frequency axis when the equivalent vibration mass decreases. After trimming, the frequency split of the resonator is reduced below 0.1 Hz, which essentially meets the functional requirements of gyroscopes.

## II.    Stiffness-Type Frequency Split Trimming

Stiffness-type frequency split trimming is performed by making a groove in the resonant ring to change the resonator's local stiffness. First, the impact of the groove in the resonant ring on the local stiffness of the resonator is analyzed. The dimensional and material parameters are shown in Tables 5.6 and 5.7. A groove of $1 \times 1 \times 1$ mm$^3$ is then made in the resonant ring, and an FE analysis is conducted to evaluate the reduction in stiffness in the trimmed position (total stiffness is $4.5 \times 10^6$ Pa). The results are shown in Fig. 5.13. When there is one groove in the resonant ring, the stiffness significantly decreases along the direction of the groove. As shown in Fig. 5.13a, the stiffness decreases at the groove by 8800 Pa, and the decreasing quantity accounts for about 2% of the total stiffness. When there are four grooves in the resonant ring, the stiffness distribution appears in a four-piece pattern. As shown in Fig. 5.13b, the stiffness of the resonator decreases at the grooves by about 4.5%. This shows that after trimming, the local stiffness of the resonator significantly decreases, and the stiffness distribution is closely related to the position of the trimming grooves. Stiffness is an important factor that affects the resonant frequency of the resonator.

| Geometric parameters | Value |
|---|---|
| Resonator radius | 12.5 mm |
| Thickness and height of the resonant ring | 1 and 8 mm |
| Bottom thickness | 0.3 mm |

**Table 5.6** Geometric dimension parameters of the resonator

**Table 5.7** Material parameters of the resonator

| Material parameters | Value | Material parameters | Value |
|---|---|---|---|
| Young's modulus, $E$ | 210 Gpa | Thermal expansion coefficient, $\gamma$ | $8.5 \times 10^{-6}$ °C$^{-1}$ |
| Alloy density, $\rho$ | 7800 kg/m$^3$ | Thermal conductivity, $\kappa$ | 60 W/m K |
| Poisson's ratio, $\mu$ | 0.3 | Typical size of thermal distribution, $\Psi$ | $1 \times 10^{-6}$ m |
| Air density, $\rho_g$ | 1 kg/m$^3$ | Specific heat per unit volume, $c$ | $3.6 \times 10^6$ J m$^{-3}$ K$^{-1}$ |
| Thickness of surface damage layer, $h_{dam}$ | 1 μm | Environmental temperature, $\Gamma$ | 293 K |

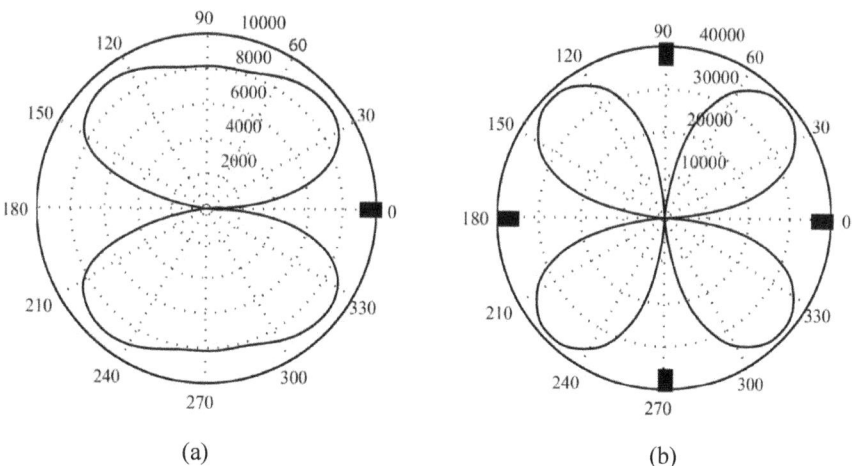

(a)          (b)

**Fig. 5.13** Stiffness changes caused by trimming. **a** Stiffness distribution of the resonator under single-groove etching; **b** stiffness distribution of the resonator under four-groove etching

Therefore, optimizing the groove structure can significantly change the frequency split.

For comparison purposes, the stiffness changes caused by mass-type trimming are also analyzed here. The mass is removed near the centerline of the resonant ring, and the hole-type trimming method shown in Fig. 5.13 is adopted. When the hole diameter is 0.75 mm and its depth is 1 mm, a frequency split of 5 Hz can be removed, and only a stiffness change of 0.18% is produced. If frequency splits of the same size are removed by the stiffness-type trimming method, a stiffness change of up to 2% will be caused by the trimming grooves. This shows that the circumferential stiffness change caused by mass-type trimming is much smaller than that caused by stiffness-type trimming.

In stiffness-type trimming, it is still assumed that the frequency split is caused by uneven mass distribution. According to the theory outlined in Sect. 5.3.1, the angle of modal deflection caused by the non-uniform mass is:

$$
\varphi_1 = \frac{\arctan\left\{ \dfrac{\sum\limits_{i=1}^{n} m_i[U(X_i)^2 + W(X_i)^2]\sin 2n\theta_i}{\sum\limits_{i=1}^{n} m_i[U(X_i)^2 + W(X_i)^2]\cos 2n\theta_i} \right\}}{2n}
$$

$$
\varphi_2 = \frac{\arctan\left\{ \dfrac{\sum\limits_{i=1}^{n} m_i[U(X_i)^2 + W(X_i)^2]\sin 2n\theta_i}{\sum\limits_{i=1}^{n} m_i[U(X_i)^2 + W(X_i)^2]\cos 2n\theta_i} \right\}}{2n} + \frac{\pi}{2n} \tag{5.13}
$$

The mode shape of the resonator is:

$$
\begin{cases}
u = U(x)\cos 2(\theta - \varphi)\cos(\omega t) \\
v = V(x)\sin 2(\theta - \varphi)\cos(\omega t) \\
w = W(x)\cos 2(\theta - \varphi)\cos(\omega t)
\end{cases} \tag{5.14}
$$

According to the principle of stiffness-type trimming, a groove can be made on the modal axis with a higher frequency in order to eliminate the frequency split. This non-ideal mass adds the following kinetic energy to the resonator:

$$
\Delta T_s = \sum_{i=1}^{n} \frac{1}{2} m_i (\dot{u}_i^2 + \dot{v}_i^2 + \dot{w}_i^2) \tag{5.15}
$$

Groove-type trimming reduces the kinetic energy of the resonator, as shown below:

$$
\Delta T_b = \sum_{i=1}^{n} \frac{1}{2} \Phi_i l_i \rho R h_r (\dot{u}_i^2 + \dot{v}_i^2 + \dot{w}_i^2) \tag{5.16}
$$

Assuming that a four-point trimming method is adopted, with groove angle width of $\Phi_i$ and a groove depth of $l_i$, therefore the reduction of the resonator's elastic potential energy is:

$$
\begin{aligned}
\Delta S = \sum_{i=1}^{n} \frac{D_r}{2R} \int_{\varphi - \Phi_i/2}^{\varphi + \Phi_i/2} \int_{l-l_i}^{l} & \left[ R^2 \left( \frac{\partial^2 w}{\partial x^2} \right)^2 + \frac{1}{R^2} \left( \frac{\partial^2 w}{\partial \theta^2} \right)^2 \right. \\
& + \frac{w^2}{R^2} + \frac{2w}{R^2} \frac{\partial^2 w}{\partial \theta^2} + 2\mu \frac{\partial^2 w}{\partial x^2} \frac{\partial^2 w}{\partial \theta^2} \\
& + 2(1-\mu) \left( \frac{\partial^2 w}{\partial x \partial \theta} \right)^2 + \frac{(1-\mu)}{2R^3} \left( \frac{\partial u}{\partial \theta} \right)^2 \\
& \left. - \frac{(1-\mu)}{R} \frac{\partial u}{\partial \theta} \frac{\partial^2 w}{\partial x \partial \theta} \right] dx d\theta
\end{aligned} \tag{5.17}
$$

After trimming, the resonant frequency of the resonator is:

$$\omega_{sp}^2 = \frac{S - \Delta S}{\overline{T} + \Delta \overline{T}_s - \Delta \overline{T}_b}$$

$$= \left(\omega_0^2 - \frac{\Delta S}{\overline{T}}\right)\left(1 - \frac{\Delta \overline{T}_b - \Delta \overline{T}}{\overline{T}}\right)^{-1} \tag{5.18}$$

where

$$\overline{T} = \frac{\rho R h_r}{2} \int_{ll}^{L+l} \int_0^{2\pi} [[U(x)^2 + W(x)^2] \cos^2\theta dx d\theta$$

$$+ \frac{\rho R h_s}{2} \int_0^l \int_0^{2\pi} [U(x)^2 + W(x)^2] \cos^2\theta dx d\theta$$

$$+ \frac{\rho h_b}{2} \int_0^{2\pi} \int_{r_0}^R U_b(r)^2 \cos^2\theta r dr d\theta$$

$$\Delta \overline{T}_b = \sum_{i=1}^n \frac{1}{2}\Phi_i l_i \rho R h_r [U(X_i)^2 + W(X_i)^2] \cos^2 2(\theta_i - \varphi)$$

$$\Delta \overline{T} = \sum_{i=1}^n \frac{1}{2} m_i [U(X_i)^2 + W(X_i)^2] \cos^2 2(\theta_i - \varphi) \tag{5.19}$$

After stiffness-type trimming, the frequency split is eliminated, therefore:

$$\omega_{sp}(\varphi = \varphi_1) = \omega_{sp}(\varphi = \varphi_2) \tag{5.20}$$

So, for a fixed value of $\Phi_i$, $l_i$ can be evaluated. This means that if the groove width is determined, the groove height can be changed to remove the frequency split, and vice versa.

Figure 5.14 shows an experimental process of stiffness-type trimming. It can be seen that stiffness-type trimming causes a continuous decrease in the overall frequency of the resonator, but since the stiffness and frequency of the high-frequency axis decrease faster, frequency matching can ultimately be realized between the drive mode and the sense mode.

### 5.3.3 Mode Trimming of Imperfect Resonators

When the driving frequency is close to the resonator's natural modal frequency, the amplitude $A_1$ of the resonator will be amplified because of the resonance effect:

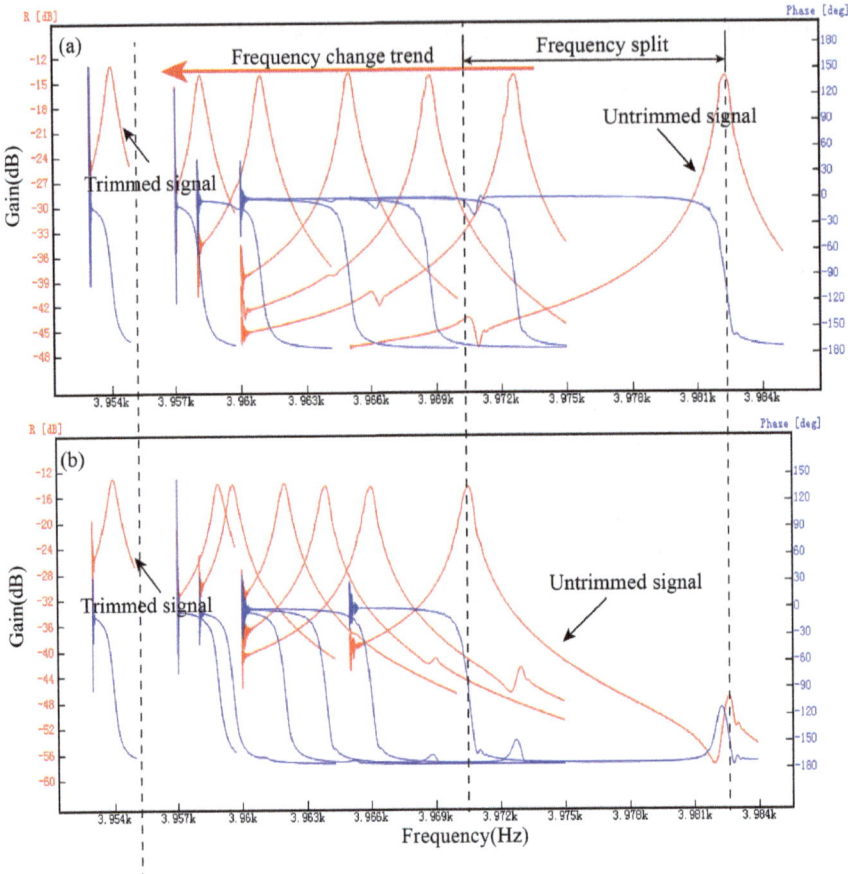

**Fig. 5.14** Stiffness-type frequency split trimming. **a** Frequency response in drive mode, **b** Frequency response in sense mode

$$A_1 = \frac{1}{\sqrt{(1 - \gamma_1^2)^2 + 4\xi^2\gamma_1^2}}A(\delta) \tag{5.21}$$

where $\gamma_1 = \omega/\omega_1$, $\omega$ represents the driving frequency; and $\omega_1$ represents the natural frequency of the drive mode.

Due to the existence of the frequency split, the sense mode is also accordingly excited, and the amplitude $A_2$ can be expressed as

$$A_2 = \frac{1}{\sqrt{(1 - \gamma_2^2)^2 + 4\xi^2\gamma_2^2}}A(\delta + 45°) \tag{5.22}$$

where $\gamma_2 = \omega/\omega_2$, $\omega_2$ represents the natural frequency of the sense mode.

Suppose an excitation vibration is a cosine vibration. Therefore, the two trains of standing waves formed in the resonator are

$$v_1(\theta, t) = A_1 \cos 2(\theta - \delta) \cos(\omega_1 t + \phi_1)$$
$$v_2(\theta, t) = A_2 \sin 2(\theta - \delta) \cos(\omega_1 t + \phi_2) \tag{5.23}$$

where $\phi_1$ and $\phi_2$ are the corresponding phase angles:

$$\phi_1 = \arctan \frac{2\xi \gamma_1}{1 - \gamma_1^2}, \phi_2 = \arctan \frac{2\xi \gamma_2}{1 - \gamma_2^2} \tag{5.24}$$

Actually, the two trains of standing waves will form a new standing wave. In this way, when the resonator is excited at the natural frequency $\omega_1$ of its drive mode:

$$v(\theta, t) = A_1 \cos 2(\theta - \delta) \cos(\omega_1 t + \phi_1) + A_2 \sin 2(\theta - \delta) \cos(\omega_1 t + \phi_2)$$
$$= \sqrt{A_1^2 + A_2^2 + 2A_1 A_2 \sin(\phi_2 - \phi_1)} \cos 2(\theta - \delta - \Theta) \cos(\omega_1 t + \phi_1)$$
$$+ A_2 \sin(\phi_2 - \phi_1) \cos[\omega_1 t + 2(\theta - \delta) + \phi_1] \tag{5.25}$$

where $\Theta = 45° - \frac{1}{2} \arctan \frac{A_1 + A_2 \sin(\phi_2 - \phi_1)}{A_2 \cos(\phi_2 - \phi_1)}$.

The first term in Eq. (5.25) represents the fact that the amplitude of the resultant standing wave is codetermined by the two trains of standing waves. The term $\delta + \Theta$ represents the azimuth of the resultant standing wave. It is related to the excitation frequency; in other words, an unstable excitation frequency destabilizes the standing wave azimuth. The second term represents the traveling wave component determined by the standing wave component of the detection axis. This traveling wave component reflects the degree of frequency mismatch within the resonator.

As can be seen, the modal deflection angle of the resonator is mainly caused by mass imbalance. Mode shape trimming requires that the standing wave loop axis be in the same direction of the driving electrode in order to reduce detection signal errors. Therefore, the modal deflection angle needs to be adjusted by mass adjustment.

Because the phase angle $\phi_2 - \phi_1$ is determined by the frequency split, the frequency split needs to be adjusted to influence the mode shape.

Suppose the resonator has a frequency split of 10 Hz at an angle of 22.5° to the driving direction, and that its corresponding modal deflection angle is also 22.5°. The mode shape can then be trimmed by reducing the resonant frequency of the resonator. The results are shown in Fig. 5.15, in which it shows that the frequency split is reduced, the modal deflection angle of the resonator decreases as well. When the frequency split is reduced below 0.3 Hz, the mode shape of the resonator coincides precisely with the drive shaft. It is noted that the relationship between the frequency split and the modal deflection angle is somewhat unsatisfactory at present. This is because as the mass imbalance decreases, the driving force plays an increasing role in enabling the azimuth of mode shape to be consistent with the drive position.

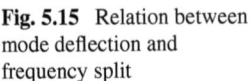

**Fig. 5.15** Relation between mode deflection and frequency split

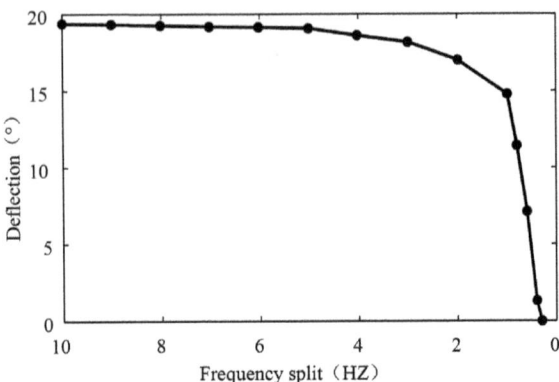

Therefore, in order to adjust the modal deflection angle to an ideal position, the resonator is only allowed to have a very small frequency split.

## 5.4   Assembly of a CVG

The gyroscope consists of a resonator and its lead package structure. The cylindrical metal shell structure is the resonator's high-precision vibration element, and must be manufactured through machining. The piezoelectric electrodes are the resonator's driving and detection units, and are fixed on the bottom of the resonator's cup-shaped metal structure by adhesive bonding techniques. Machined resonators must be fixed on a specific mounting base and connected to the lead circuit board through a lead wire. Because manufacturing errors are inevitable in the resonator's preci-sion machining process and in the process of bonding the piezoelectric electrodes, the resonator must be mechanically balanced in order to protect itself from being disturbed by external air currents while it is working. As a result, it must be enclosed in a sealed package.

The manufacturing and assembly of a CVG consists of the following several steps:

(1)   Precision machining of the cylindrical metal shell structure;
(2)   Bonding and connection of piezoelectric electrodes;
(3)   Assembly of the resonator, lead circuit board, and mounting base;
(4)   Mechanical balance of the resonator;
(5)   Encapsulation of the resonator.

The piezoelectric electrodes are the gyroscope's driving and sensing elements. They are also the key to its precision assembly. The PZT-5 piezoelectric foil gauges, which have a high piezoelectric constant, are selected as the piezoelectric elec-trodes used for both drive and detection. At present, the most common method for connecting piezoelectric foil gauges to an actuator or sensor is bonding. Moreover, the method of bonding connecting piezoelectric electrodes to a metal elastomer has been successfully applied to the R&D of ultraacoustic motors. This bonding technique should ensure that the resonator's piezoelectric electrodes are positioned accurately, and that each bonding layer is uniformly thick and defect-free.

There is no locating plane on the outer bottom of the resonator's metal structure that can be used for the bonding of piezoelectric electrodes, but if the piezoelectric electrodes are directly bonded to the outer bottom of the metal structure, locating errors will be caused. A bonding fixture as shown in Fig. 5.16 can be designed for assembly. The locating base consists of a piezoelectric electrode locating slot and a preload spring. The depth of the piezoelectric electrode locating slot is smaller than the thickness of the piezoelectric electrodes. The locating accuracy of the piezoelectric electrodes on the metal structure is ensured through the precision of the fixture's machining.

The piezoelectric electrode bonding scheme consists of the following processing steps:

(1)  The eight piezoelectric electrodes are to be spin-coated with adhesive, and the residual adhesive should be cleared from the edges of the piezoelectric electrodes.

(2)  The piezoelectric electrodes are to be in the locating slot in the locating base, and glued face-up.

**Fig. 5.16** Bonding fixture and process flow diagram

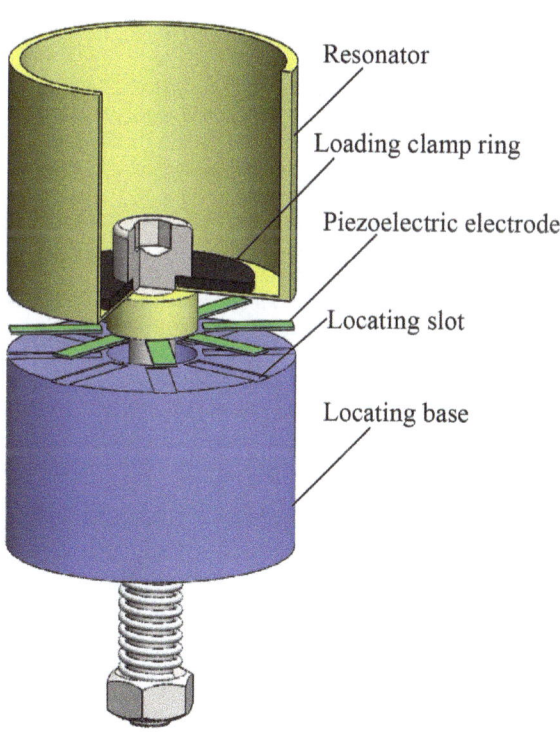

Resonator

Loading clamp ring

Piezoelectric electrode

Locating slot

Locating base

(3)    The loading clamp ring is to be placed in the inner bottom of the resonator's metal structure.

(4)    The metal structure of the resonator should be inserted into the center hole in the locating base through the support bar, and the locating pin should then be inserted into the loading board, the metal structure of the resonator, and the locating hole in the locating base in their proper order. The metal structure of the resonator should then be gently placed on the glued faces of the eight piezoelectric electrodes.

(5)    The screw should be inserted into the metal structure of the resonator and the center hole of the locating base. The nut on the screw imposes a pre-loaded pressure on the loading board through the spring. This pressure exists between the resonator's metal structure and the piezoelectric electrodes. The loading clamp ring ensures that all piezoelectric electrodes are stressed uniformly.

(6)    The pre-compressed resonator and bonding fixture should be placed in a temperature control box to have the bonding layer dried at the set temperature within the set time.

# References

1. Fumin, C., Guojun, L., Deda, S. (1986). *Elastic alloys*. Shanghai: Shanghai Science and Technology Press
2. Corporation SM. (2004). NI-SPAN-C alloy 902 Specifications
3. Zidan, W. (2016). *Analysis of the heterogeneity and micro-deformation of the CVG resonator*. Changsha: National University of Defense Technology.
4. Heat Treatment Society, Chinese Mechanical Engineering Society. (2008). *Handbook of heat treatment*. Beijing: Machine Press
5. Yi, T. (2011). *Research on the key technology of cup-shaped wave gyros*. Changsha: National University of Defense Technology.
6. Fox, C. H. J. (1990). A simple theory for the analysis and correction of frequency splitting in slightly imperfect rings. *Journal of Sound and Vibration, 142*(2), 227–243.
7. Choi, S. Y., & Kim, J. H. (2011). Natural frequency split estimation for inextensional vibration of imperfect hemispherical shell. *Journal of Sound and Vibration, 330*(9), 2094–2106.
8. Xiang, Xi. (2014). *Research on the drift mechanism and suppression technology of cup-shaped wave gyroscopes*. Changsha: National University of Defense Technology.

# Chapter 6
# Parameter Testing Methods for CVG Resonators

The performance of a gyroscope's resonator is a decisive factor affecting its overall performance. Therefore, when the performance parameters of a resonator are tested, special attention should be paid to the following performance parameters: the mechanical quality factor, frequency split, and the position of the rigid axis.

The mechanical quality factor, called the $Q$ value, refers to the ratio of the total energy stored in a vibration system to the energy lost during a vibration period. When the resonator works, the higher the quality factor, the lower the energy loss. Increasing the quality factor of the resonator helps to reduce the output errors and energy consumption of the CVG resonator and improve its sensitivity. A high $Q$ value helps to improve the sensitivity of the resonator. The more uniform the $Q$ value is, the more stable the resonator's vibrations. To improve the resonator's overall performance, we must select a suitable material and machining and thermal treatment techniques in order to produce a very high $Q$ value.

The frequency split refers to a difference in frequency between the resonator's drive mode and sense mode [1, 2]. An ideal resonator has no frequency split, and when it is excited, the resonator will only vibrate in its drive mode. However, owing to machining errors, poor material properties, and many other factors in the actual machining process, the resonator will have two natural rigid axes, and these are at an angle of 45° to each other. The resonator has different natural frequencies along the two axes, producing orthogonal errors and causing gyroscopic drift.

## 6.1 Methods for Testing Resonator Frequency Characteristics

The resonant frequency is an important parameter for the operation and testing of the resonator. When the drive signal frequency is the same as the resonant frequency of the resonator, it can excite the drive mode of the resonator, allowing the vibration of the resonator to be measured. The resonator is made of an iron-based alloy. An

© National Defense Industry Press 2021

X. Wu et al., *Cylindrical Vibratory Gyroscope*, Springer Tracts in Mechanical Engineering,
https://doi.org/10.1007/978-981-16-2726-2_6

**Fig. 6.1** Diagram of
electromagnetic drive system

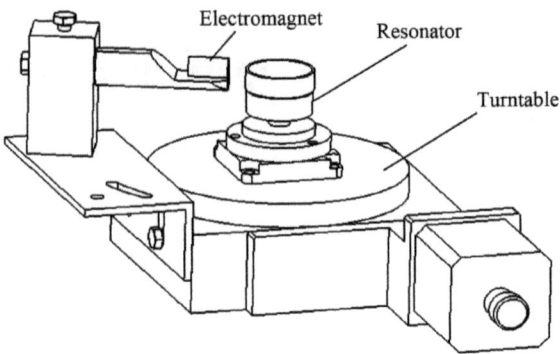

electromagnet can be used to apply an alternating force on the resonator in order to
make it vibrate in its drive mode. The magnitude of the magnetic attraction to the
resonator depends on its operating range and driving voltage. Alternating exciting
currents can be applied to the electromagnet to exert an alternating attractive force
on the resonator to perform non-contact excitation of vibration [3] (Fig. 6.1).

Figure 6.2 shows a vibration signal detection system. A signal detection device
with a MEMS microphone enables the detection position of the MEMS microphone
to move around the circumference of the resonator by using a turntable, allowing the
distance between the MEMS microphone and the resonator shell wall to be adjusted,
and the position of the MEMS microphone to be circumferentially fine-tuned without
moving the resonator.

A resonator without piezoelectric electrodes is chosen as a test object and fixed on
the turntable to drive and detect it in different positions. A drive position is selected at
random, and a frequency response analyzer is used to excite the resonator and collect
the signals detected by the MEMS microphone in order to trace the gain-frequency
curve and phase-frequency curve. If the drive position is not a rigid vibration axis
position, the gain-frequency response curve should contain two peaks (Fig. 6.3),
and the frequencies corresponding to these peaks are the resonant frequencies of the
resonator in different rigid axis directions.

**Fig. 6.2** Schematic of an
acoustic detector in a
resonator

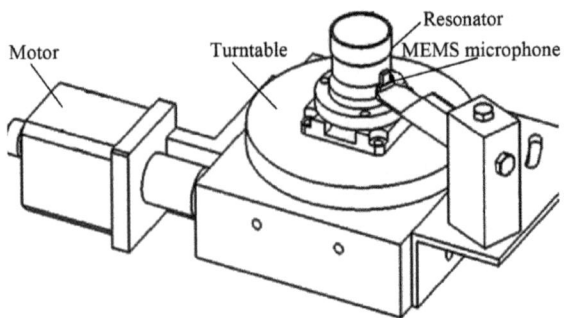

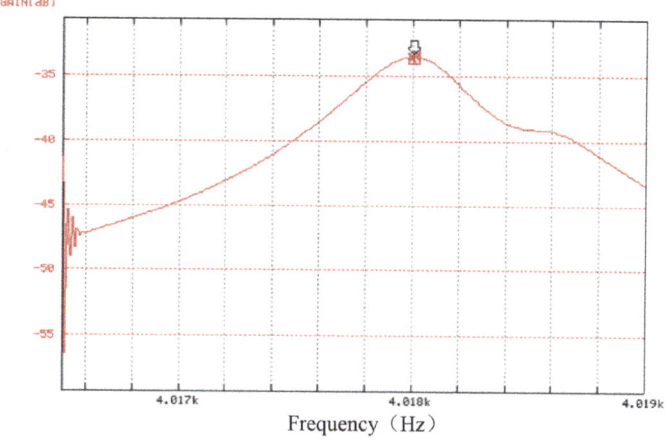

**Fig. 6.3**  Frequency sweep results at non-rigid axis positions

As can be seen from Fig. 6.3, the gain-frequency curve has two peak points, the frequencies of which are 4018.10 Hz and 4018.53 Hz respectively, and these correspond to the resonant frequencies of the resonator's two rigid axes.

## 6.2  Methods for Testing Resonator Quality Factors

The mechanical quality factor refers to the ratio of the total energy stored in the system to the energy lost in a certain period. It can be calculated using the $-3$ dB method:

$$Q = \frac{f^*}{\Delta f} \tag{6.1}$$

where $f^*$ represents the resonant frequency of the resonator; and $\Delta f$ represents the $-3$ dB bandwidth of the resonator, referring to a difference between two frequencies that rely on the resonant frequency of the resonator as a standard, and have a gain of $-3$ dB.

Figure 6.4 shows the amplitude-frequency curve of the first rigid axial position of the resonator under electromagnetic actuation.

The data required for $Q$ value calculation are separated from the data recorded by the frequency response analyzer. The results of the frequency sweep show that the center frequency is 4018.54 Hz, and the $-3$ dB bandwidth is 440.9 MHz. These data are then substituted into Eq. (6.1) to calculate the $Q$ value, which is equal to 9133.05.

Figure 6.5 shows the frequency sweep results at the second rigid axial position achieved through non-contact actuation.

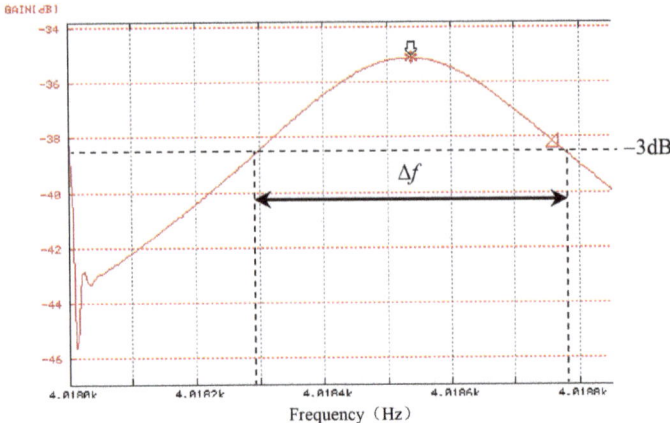

**Fig. 6.4** Frequency sweep curve of the first rigid axis drive and wave loop position

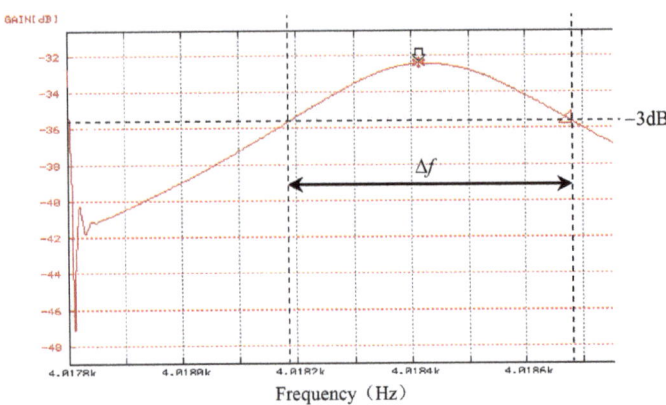

**Fig. 6.5** Frequency sweep curve at the second axial position and wave loop position

The frequency sweep results at this rigid axial position show that the center frequency is 4018.42 Hz, its −3 dB bandwidth is 472.5 MHz, and its $Q$ value is 8513.58.

To compare the test results of the $Q$ value in different driving modes, we chose to perform frequency sweeps in the two rigid axial positions under piezoelectric actuation and then record the data. The data are then compared with the measured $Q$ value. The measured $Q$ values of the first and second rigid axes are 8930.47 and 8201.22 respectively.

The $Q$ value of each rigid axis can be shown to decrease to some extent after the piezoelectric electrodes are bonded to the resonator. This is mainly because the piezoelectric electrodes and bond increase the internal friction of the resonator during its vibrations, affecting the $Q$ value of the resonator. The difference in the $Q$

value between the two rigid axes is caused by machining errors and inhomogeneous resonator materials.

## 6.3 Methods for Testing Resonator Mode Shape

### I. Direct Measurement Method

In an ideal resonator, the fourth and fifth-order modal frequencies are the same and its mode shapes also appear in the corresponding excitation positions. However, in imperfect resonators, due to inconsistencies in stiffness and mass in various directions, not is a frequency split present, its mode shapes do not necessarily operate in the corresponding piezoelectric driving directions.

An experiment can be conducted to directly observe the vibration of the resonator at its operating frequencies. The vibration measurer (vibrometer) is a laser scanning and measurement instrument that consists of a scanning-based optical head, a controller, a connection box, and a data management system. After a measurement area and measurement points are arbitrarily defined within a target video image, a scanning measurement will be automatically performed under the control of the data management system.

The amplitude measurement system is shown in Fig. 6.6. The test piece is a Ø25 mm prototype CVG resonator. The frequency response analyzer shows that the fourth-order modal frequency of the resonator is 4440 Hz, while the fifth-order modal frequency is 4441.5 Hz. 5 V harmonic excitation is then applied to a set of piezoelectric electrodes at a certain frequency in order to resonate the resonator.

The laser vibrometer can measure a number of points on the same generatrix of the resonating ring at the same time. The turntable can then be rotated to measure the radial amplitude of each point on the circumference of the resonant ring. 3 evenly spaced points on the same generatrix of the resonant ring are then measured. The turntable is then rotated to measure a set of amplitudes at every 5°. Figure 6.7 shows the polar coordinates of the amplitudes (μm) in the fourth and fifth-order modes. Figure 6.7a shows the fourth-order mode under 4440 Hz excitation, while Fig. 6.7b shows the fifth-order mode under 4441.5 Hz excitation. The test points on the generatrix are shown in the figure from inside-out and from low to high.

The test results in Fig. 6.7 show that, when the resonator is excited by the piezoelectric electrodes, the corresponding mode shapes appear at the fourth-order modal frequency and the fifth-order modal frequency, and the two mode shapes are at angle

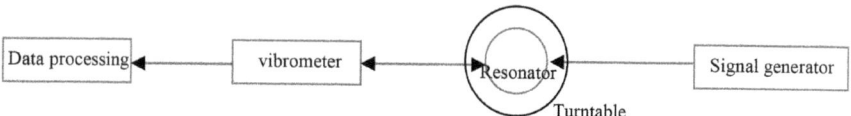

**Fig. 6.6** Diagram of the amplitude measurement system

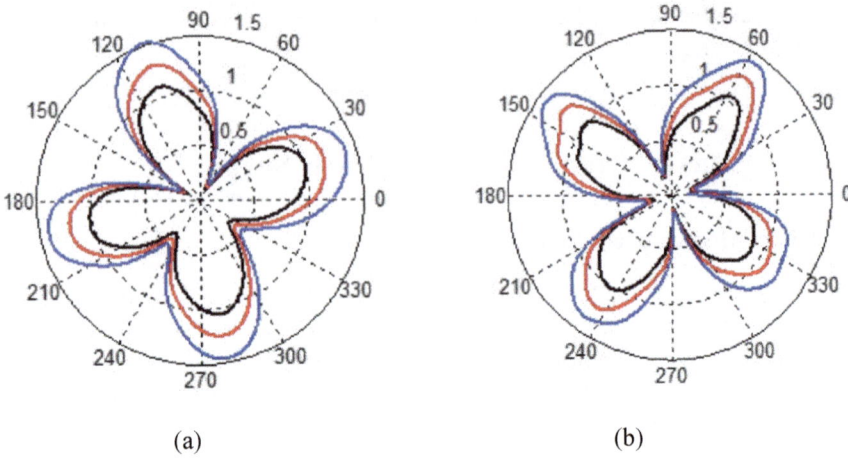

**Fig. 6.7** Measured resonator modes **a** fourth-order mode; **b** fifth-order mode

of 45° to each other. The amplitude of the resonator is subject to a near-sinusoidal variation. The maximum amplitude of the CVG resonator is about 1.5 μm, while the minimum amplitude appears in the node location.

The exciting piezoelectric electrode is located in the direction of 0°, and its ideal mode shape is shown in Fig. 6.8. A comparison of the two figures shows that the measured and theoretical mode shapes are basically identical, but their directions are different. Theoretically, the excitation occurs in the same direction as the mode shape, but in actuality, there is an angle of deflection between the excitation and the mode shape. This is because the direction of the resonator's resonation follows a rigid

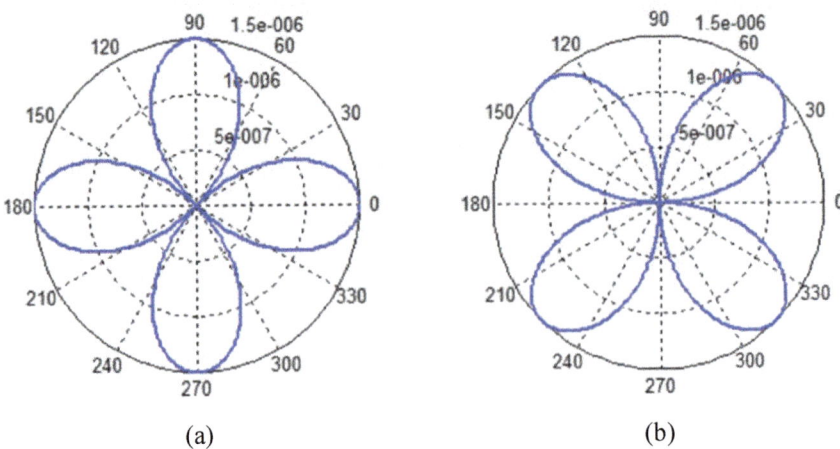

**Fig. 6.8** Theoretical mode shapes of the resonator **a** the fourth-order mode; **b** the fifth-order mode

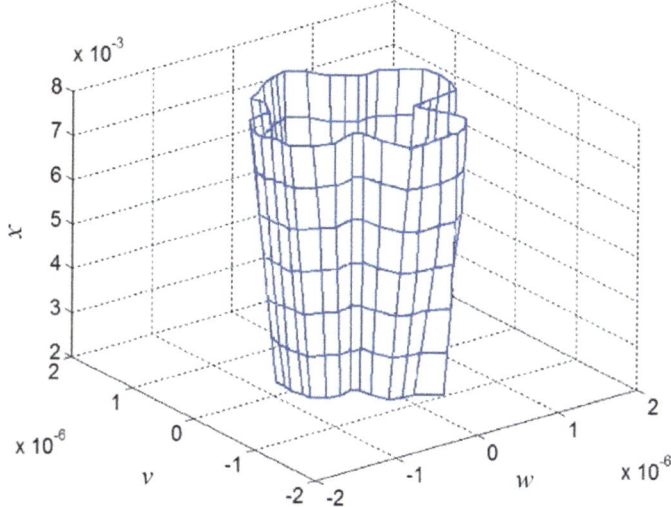

**Fig. 6.9**  The overall radial amplitude distribution of the resonant ring

axis with a phase difference of 45°, while the location of the rigid axis is an inherent attribute of the resonator and does not change with excitation. The rigid axis of an imperfect resonator may be inconsistent with its excitation, so the modal direction is different from the excitation direction as well. To sum up, the modal errors of an imperfect resonator are mainly reflected in the following characteristics:

(1)   The mode shape deviates from the direction of excitation in the piezoelectric electrodes;
(2)   The amplitude of the standing wave vibrations is unstable.

The amplitude of the resonator changes linearly in the direction of the generatrix, as shown in Fig. 6.9. If the resonant ring is measured along the generatrix, several sets of data can be obtained and used to reveal the overall radial amplitude distribution of the resonant ring.

## II.   Indirect Measurement Method

The experimental system shown in Figs. 6.1 and 6.2 is used to measure the resonator's vibration characteristics. After the resonator is fixed on the turntable, the frequency response analyzer is used to excite the electromagnet to sweep the frequency of the resonator at every 5° within a 90° range. Because the frequency sweep results contain two peaks when the resonator is driven in a non-rigid axial location, the closer the drive position is to a certain rigid axial location, the more greatly the two peak positions differ from each other in amplitude. When the sweep results contain only one peak position, this peak position can be considered to be near a rigid axial location on the resonator.

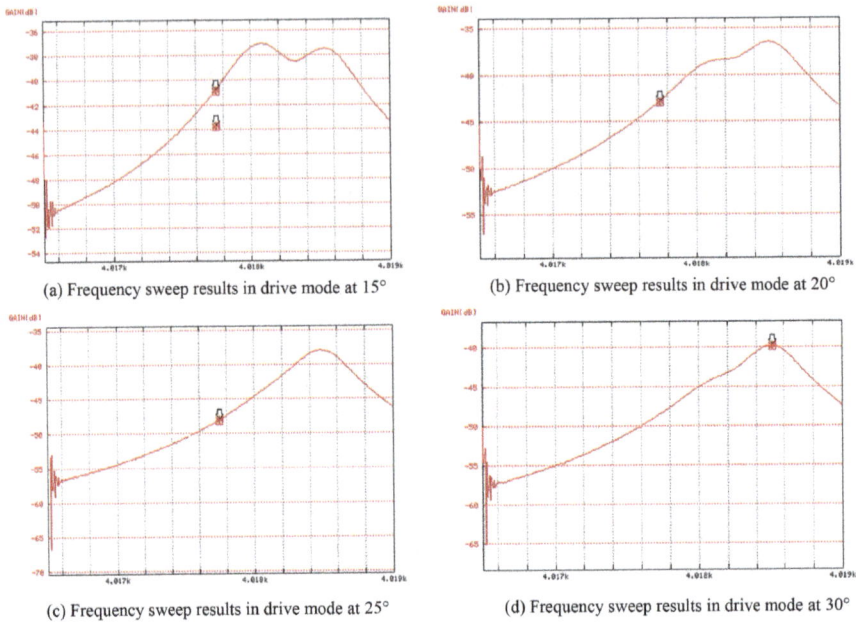

(a) Frequency sweep results in drive mode at 15°      (b) Frequency sweep results in drive mode at 20°

(c) Frequency sweep results in drive mode at 25°      (d) Frequency sweep results in drive mode at 30°

**Fig. 6.10** Frequency sweep results at different drive positions near 25°

To verify the accuracy of rigid axial position detection, a piezoelectric electrode can be bonded to a rigid axial position on the resonator after it is detected. In addition, the frequency response analyzer should be used to perform frequency sweep measurements of the accuracy of the rigid axial position. Drive positions are taken at every 5° during the frequency sweep. These results are shown in Fig. 6.10.

It can be seen from Fig. 6.10 that a single-peak curve is traced 25°, while another harmonic peak can be observed at 20° or 30°. According to the definition of the rigid axis, there is a rigid axis near 25°.

As in the previously mentioned 25° position, a single-peak curve is traced at 70°, while other harmonic peaks can be observed at 65° and 75°. There can thus be considered to be a rigid axis in the resonator near 70° (Fig. 6.11).

To locate a rigid axis more accurately, the MEMS microphone is used to detect the drive positions by 45° intervals within the range of 20°–30° and 65°–75°. The detection outputs of the MEMS microphone are compared in different drive positions. The recorded data are shown in Tables 6.1 and 6.2.

From the above data, the detection amplitude of the MEMS microphone can be seen to reach a minimum at 26° and 71°. According to the definition of the rigid axis of vibration, the two drive positions are the rigid axes of the resonator's vibration.

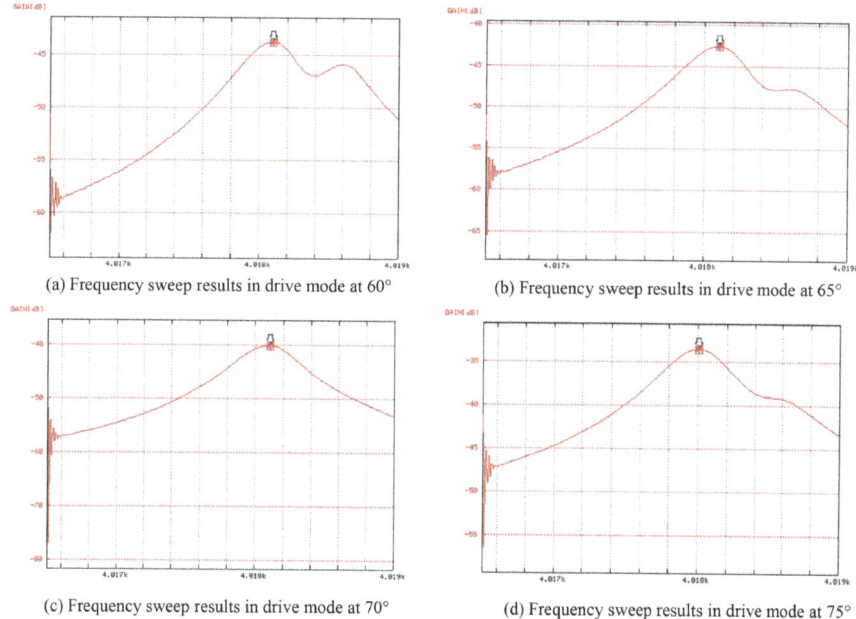

(a) Frequency sweep results in drive mode at 60°

(b) Frequency sweep results in drive mode at 65°

(c) Frequency sweep results in drive mode at 70°

(d) Frequency sweep results in drive mode at 75°

**Fig. 6.11** Frequency sweep results near 75° drive positions

**Table 6.1** Detection results near 25°

| Drive position (°) | 21 | 22 | 23 | 24 | 25 |
|---|---|---|---|---|---|
| Detection data (mV) | 74.7 | 74.9 | 75.0 | 74.8 | 74.2 |
| Drive position (°) | 26 | 27 | 28 | 29 | 30 |
| Detection data (mV) | 71.9 | 72.0 | 74.5 | 78.1 | 79.0 |

**Table 6.2** Detection results near 70°

| Drive position (°) | 66 | 67 | 68 | 69 | 70 |
|---|---|---|---|---|---|
| Detection data (mV) | 87.1 | 86.0 | 85.6 | 82.0 | 80.5 |
| Drive position (°) | 71 | 72 | 73 | 74 | 75 |
| Detection data (mV) | 80.4 | 82.9 | 83.3 | 85.9 | 86.6 |

## 6.4 Methods for Testing the Parameters of Piezoelectric Electrodes

As shown in Fig. 6.12, the electrode face is on a principal plane perpendicular to the $z$-axis and is polarized in the direction of its thickness. The electric field is applied along the direction of thickness. The length $l$ is along the $x$-axis, the width $a$ is along the $y$-axis, and the thickness $t$ is long the $z$-axis. The length axis is the main factor,

**Fig. 6.12** Spatial structure
of the piezoelectric vibrator

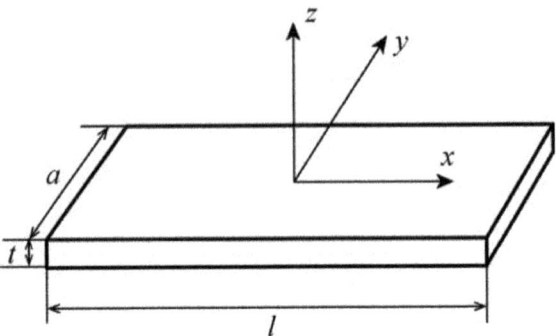

so only the effect of the stress component $X_1$ along the x-axis is considered. Also, because the electrode face is perpendicular to the z-axis, only the effect of the electric field component $E_3$ is considered. $X_1$ and $E_3$ are selected as independent variables, and the first type of piezoelectric equation is selected:

$$\begin{cases} x_1 = s_{11}^E X_1 + d_{31} E_3 \\ D_3 = d_{31} X_1 + \varepsilon_{33}^X E_3 \end{cases} \tag{6.2}$$

where $s_{11}^E$ is an elastic compliance constant; $\varepsilon_{33}^X$ is a free dielectric constant; $x_1$ represents the stress along the x-axis; and $D_3$ represents the electric displacement along the z-axis.

In the long and thin piezoelectric ceramic vibrator shown in Fig. 6.12, the direction of the vibrator's polarization is parallel to the direction of thickness, the face of the electrode is perpendicular to the direction of its thickness, and both ends of the piezoelectric electrode are a mechanically free state. Under the action of the applied alternating electric field $E_3 = E_0 e^{j\omega t}$, the long and thin piezoelectric ceramic vibrator generates stretching vibrations along the direction of its length. The direction of vibration at each point in the vibrating body and the direction of vibration propagation follow the length axes of the long and thin piezoelectric ceramic vibrator.

Based on Newton's second law of motion, a motion equation is established for the long and thin piezoelectric ceramic vibrator:

$$\rho \frac{\partial^2 u}{\partial t^2} = \frac{\partial X_1}{\partial x} \tag{6.3}$$

The piezoelectric electrode face is an isopotential surface, and the electric field component $E_3$ is uniformly distributed throughout the wafer, so $\frac{\partial E}{\partial x} = 0$. The following wave equation can be built for the piezoelectric electrodes based on Eqs. (6.2) and (6.3):

$$\frac{\partial^2 u}{\partial t^2} = \frac{1}{\rho s_{11}^E} \frac{\partial^2 u}{\partial x^2} = c^2 \frac{\partial^2 u}{\partial x^2} \tag{6.4}$$

Longitudinal vibrations are generated by piezoelectric effect $d_{31}$, so the general solution of Eq. (6.4) is as follows:

$$u = [A \cos(kx) + B \sin(kx)]e^{j\omega t} \tag{6.5}$$

where $k = \omega/c$. Equation (6.5) is substituted into Eq. (6.3), with the following equation able to be derived:

$$X_1 = \frac{1}{s_{11}^E}[-A \sin(kx) + B \cos(kx)]ke^{j\omega t} - \frac{d_{31}}{s_{11}^E}E_0 e^{j\omega t} \tag{6.6}$$

The piezoelectric vibrator has two free ends, and its mechanical free boundary condition is as follows:

When $x = 0$, $X_l|_{x=0} = 0$; when $x = l$, $X_l|_{x=l} = 0$, therefore:

$$X_1(x, t) = \frac{d_{31}E_3}{s_{11}^E}\left[\frac{\sin(k(l - x)) + \sin(kx)}{\sin(kl)} - 1\right] \tag{6.7}$$

According to piezoelectric equation (6.2), the following statement is derived:

$$D_3(x, t) = \frac{d_{31}^2 E_3}{s_{11}^E}\frac{\sin(k(l - x)) + \sin(kx)}{\sin(kl)} + \left(\varepsilon_{33}^X - \frac{d_{31}^2}{s_{11}^E}\right)E_3 \tag{6.8}$$

The current $I_3$ on the electrode surface is equal to the time-varying change rate of the charge $Q_3$, therefore:

$$I_3 = \frac{dQ_3}{dt} \tag{6.9}$$

The relationship between the charge $Q_3$ on the electrode surface and the electrode $D_3$ is as follows:

$$Q_3 = \int_0^l \int_0^a D_3 dx dy = la\left(\varepsilon_{33}^X - \frac{d_{31}^2}{s_{11}^E}\right)E_3 + \frac{d_{31}^2 E_3 la \tan\left(\frac{kl}{2}\right)}{s_{11}^E} \frac{1}{\frac{l}{2}} \tag{6.10}$$

$$I_3 = j\omega la\left\{\varepsilon_{33}^X + \left[\frac{\tan\left(\frac{kl}{2}\right)}{\frac{kl}{2}} - 1\right]\frac{d_{31}^2}{s_{11}^E}\right\}E_3 \tag{6.11}$$

When the external frequency is equal to the resonant frequency, the piezoelectric ceramic plate vibrates. When the vibrator resonates, its amplitude reaches a maximum, and so does its elastic energy. On the other hand, since the piezoelectric ceramic has a piezoelectric effect, the electrical signal input method can be used to mechanically vibrate the piezoelectric ceramic plate through the inverse piezoelectric effect. The mechanical resonance of the ceramic plate generates and outputs an electrical signal through the direct piezoelectric effect.

A long and thin plate with the dimensions of 60 mm × 2 mm × 0.2 mm is selected and connected to a PZT5 piezoelectric electrode as well as a $zx$-cut wafer with $d_{31} \neq 0$ by the method shown in Fig. 6.13. When the frequency of the signal generator rises slowly, and the signal frequency is equal to a certain value, the current transmitted through the piezoelectric ceramic vibrator reaches a maximum, but after the signal frequency changes, the current reaches a minimum. See Fig. 6.14.

As can be seen in Fig. 6.14, the current flowing through the piezoelectric ceramic vibrator changes with frequency, suggesting that the equivalent impedance of the piezoelectric ceramic vibrator also changes with frequency. When the signal frequency is equal to $f_m$, the current flowing through the piezoelectric ceramic vibrator reaches a maximum; i.e., the equivalent impedance reaches a minimum, while the admittance reaches a maximum. When the signal frequency equal to the $f_n$, current flowing through the piezoelectric ceramic vibrator reaches a minimum, the equivalent impedance reaches a maximum while the admittance reaches a

**Fig. 6.13** Diagram for piezoelectric electrode test

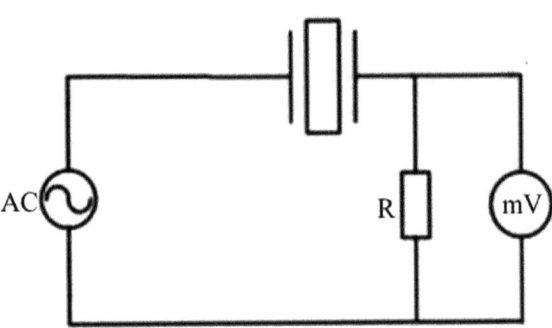

**Fig. 6.14** Current change dependent on frequency

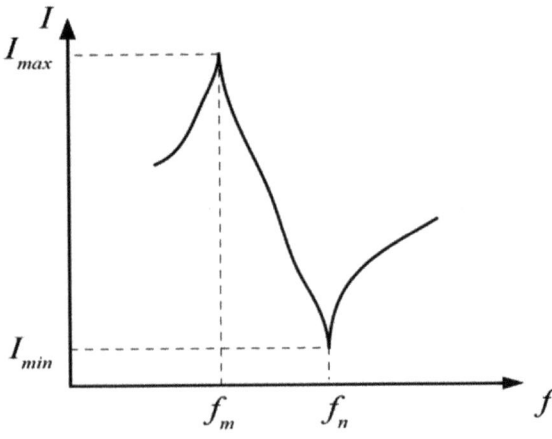

minimum. Therefore, $f_m$ is generally called the maximum admittance frequency (i.e., maximum transmission frequency) or minimum impedance frequency, while $f_n$ is called the minimum admittance frequency (i.e., minimum transmission frequency) or maximum impedance frequency.

If the signal frequency of the signal source continues to be increased, a series of sub-maximums and sub-minimums will appear regularly in the current. These frequencies respectively correspond to the resonant frequencies of other vibration modes of the piezoelectric ceramic vibrator and to the resonant frequencies of the high-order vibration modes of the piezoelectric ceramic vibrator.

Considering the effect of the temperature introduced in the piezoelectric beam model, when the frequency of the alternating voltage is equal to the resonant frequency, the piezoelectric ceramic beam will resonate. The resonant frequency $\omega_n$ of the Timoshenko beam is then introduced:

$$\omega_n = \left(\frac{n\pi}{l}\right)^2 \frac{E_p I}{\rho A} \tag{6.12}$$

where $l$ represents the length of the beam; $A$ represents the cross-sectional area of the beam; $n$ represents the order of beam vibrations; $I$ represents the cross-sectional moment of inertia; and $E_p$ represents the elastic modulus of the beam.

The temperature effect on elastic modulus of the piezoelectric ceramic:

$$E_p = E_{p0}[1 + \alpha(T - T_0)] \tag{6.13}$$

where $E_{p0}$ represents the elastic modulus of the piezoelectric ceramic plate at room temperature, equal to 139 GPa; $\alpha$ represents the temperature coefficient of the elastic modulus of the piezoelectric ceramic plate, and $T_0 = 25$ °C at room temperatures.

The electromechanical coupling coefficient $k_{31}$ of the piezoelectric ceramic can be calculated through the following formula:

$$k_{31} = \frac{d_{31}^2}{\varepsilon_{33}^X s_{11}^E} \tag{6.14}$$

where the free dielectric constant $\varepsilon_{33}^X$ has the following relationship with the capacitance $C$ of the piezoelectric ceramic plate, the electrode area $A$, and the interelectrode distance $t$: $\varepsilon = C \times t/A$.

Equation (6.14) is then substituted into Eq. (6.11), deriving the following equation:

$$I_3 = jwla\left\{\varepsilon_{33}^X + \left[\frac{\tan\left(\frac{kl}{2}\right)}{\frac{kl}{2}} - 1\right]k_{31}\varepsilon_{33}^X\right\}E_3 \tag{6.15}$$

The relationship between the elastic compliance coefficient and the resonant frequency is as follows:

$$s_{11}^E = \frac{(2n-1)^2}{4l^2 \rho \omega_n^2} \tag{6.16}$$

where $n$ represents the order of the beam mode shape.

The relationship between the piezoelectric coefficient and the amplitude is as follows:

$$d_{31} = \frac{x \cdot 2\zeta \cdot 2t}{U_0 \cdot l} \tag{6.17}$$

$\zeta$ represents the damping coefficient. The damping of the cantilever beam is little affected by temperature. The vibration displacement amplitude $x$ is calculated according to the formula $x = \dot{x}/(2\pi \omega_n)$.

**Example of Piezoelectric Ceramic Parameter Testing**

The test method is shown in Fig. 6.15 [4]. The FRA5087 frequency response analyzer is used for signal excitation and detection. The exciting voltage is a sinusoidal voltage of 5 V. In the process by which the vibration of the piezoelectric ceramic cantilever beam progresses to resonance, the larger the current $I$ flowing through resistance $R$, the larger the voltage drop at $R$. When the vibration of the cantilever beam reaches the resonance point, the current flowing through $R$ reaches a maximum, as does the voltage drop. The voltage detected at point 1 in Fig. 6.6 is the lowest, so CH1/CH2 is largest, corresponding to point $f_m$ in Fig. 6.16.

After it is placed in a temperature control box, the piezoelectric ceramic cantilever beam must be fixed with a restraining device. The experimental temperature range is set to $-40$ to $60\,°C$. The beam is tested three times at each temperature point. After performing a frequency sweep with FRA, two orders of modes can be observed on the screen, as shown in Fig. 6.16.

The same-order modal resonance points of the piezoelectric ceramic cantilever beam at all temperature points are detected via FRA. The resonant frequency value

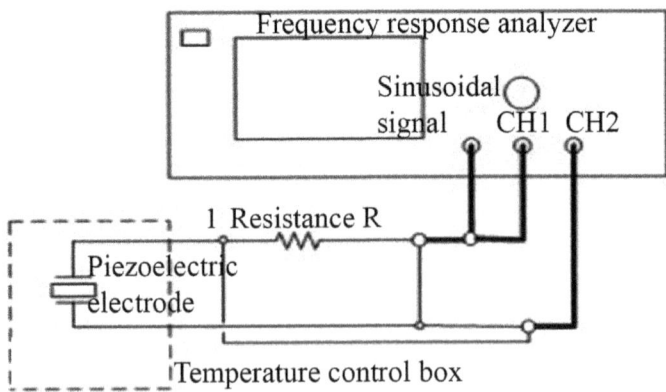

**Fig. 6.15**   Diagram of piezoelectric cantilever beam test

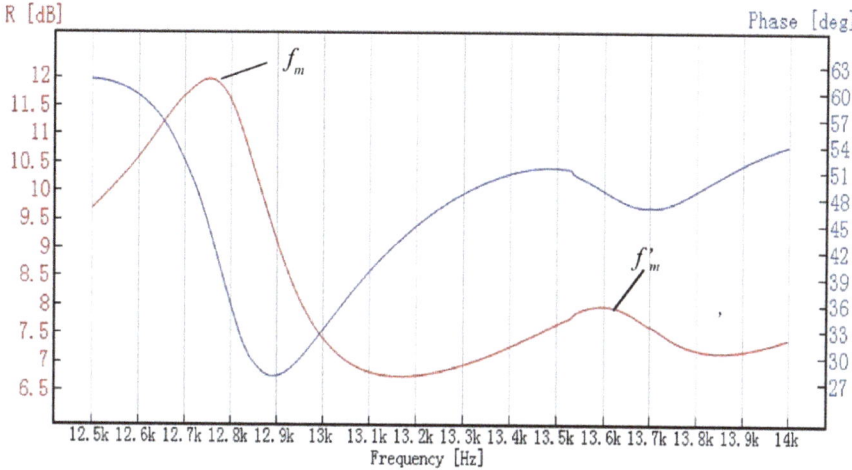

**Fig. 6.16** Resonance spectrum analysis chart

and gain value (i.e., CH1/CH2) corresponding to each resonance point $f_m$ are then recorded, as shown in Table 6.3.

A Polytec laser vibrometer is then used to measure the vibration of the piezoelectric ceramic cantilever beam. When the temperature changes, a single-point velocity test is performed on the piezoelectric ceramic cantilever beam to evaluate the single-point amplitude of the cantilever beam when the temperature changes. The measured single-point vibration velocity is shown in Fig. 6.17.

As the temperature rises from −40 to 60 °C, the amplitude of the single-point vibration displacement on the beam has a relationship with its temperature, and this relationship is shown in Fig. 6.18.

**Table 6.3** The resonant frequency and gain of the cantilever beam at different temperatures

| Temperature (°C) | Resonant frequency (Hz) | Gain at point $f_m$ (dB) |
|---|---|---|
| −40 | 12,754.7168 | 11.966 |
| −30 | 12,690.4138 | 11.761 |
| −20 | 12,640.4726 | 11.932 |
| −10 | 12,571.9208 | 11.658 |
| 0 | 12,500.0928 | 11.327 |
| 10 | 12,454.8897 | 11.235 |
| 20 | 12,380.0437 | 11.165 |
| 30 | 12,320.1376 | 11.585 |
| 40 | 12,247.6874 | 11.033 |
| 50 | 12,170.8233 | 11.350 |
| 60 | 12,094.9395 | 11.685 |

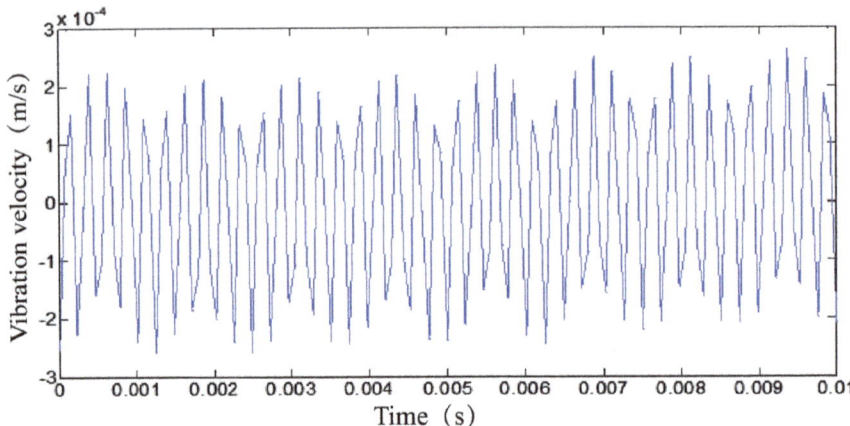

**Fig. 6.17**  Resonant velocity of the piezoelectric electrode

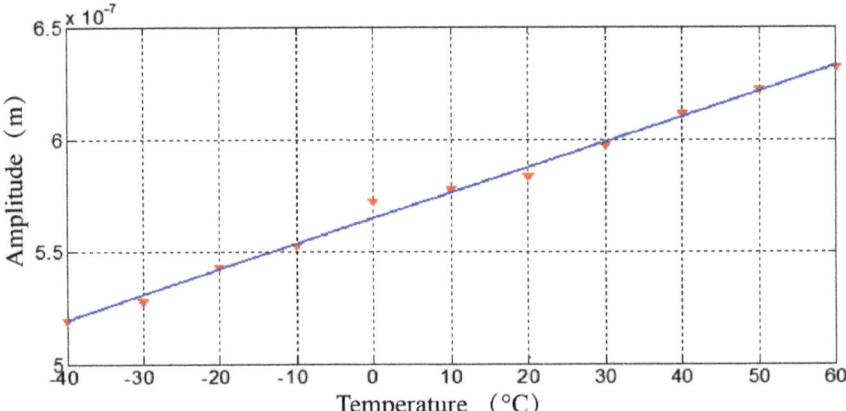

**Fig. 6.18**  Relationship between the vibration amplitude and temperature

By performing a linear fit according to Fig. 6.18, the relationship between the vibration amplitude and temperature can be found through the following equation: $y = p_1 \times x + p_2$.

where $p_1 = 1.1306 \times 10^{-9}$, $p_2 = 5.6497 \times 10^{-7}$.

It can therefore be determined that the micro-vibration of the piezoelectric electrode increases proportionately with temperature, and that the variation coefficient is $1.1306 \times 10^{-9}$ m/°C.

## 6.5   Simple Method for Testing Comprehensive Resonance Parameters

This section introduces a method for measuring the parameters of resonators based on acoustic wave superposition and decomposition [5]. It is easily observed that the acoustic pressure fields formed by structural vibrations are closely related to the shape of the vibrations, so vibration information can be collected from detected acoustic signals. The free vibration of a resonator is composed of a series of basic modes. However, only the elliptical standing wave mode can be stably maintained for extended periods, because the ring structure of the resonator easily suppresses other modes during its natural vibrations. Considering that the standing wave mode is the CVG's main working mode, the free vibration of the resonator can be used to test its parameters.

When an imperfect oscillator freely vibrates, its mode shape is composed of two standing waves with different frequencies at an angle of 45° to each other. As a result, differences in resonator parameters mean different resultant standing waves. Therefore, the parameter fitting method can theoretically be used to calculate a resonator's parameters.

For resonators vibrating in an atmosphere, the distribution of the resulting acoustic pressure field is shown in Fig. 6.19. As can be seen, the highest intensity of acoustic pressure corresponds to the antinode of the standing wave and the lowest intensity of acoustic pressure corresponds to the nodal position. Moreover, the direction of the acoustic pressure gradient indicates the direction of the mode shape.

A mode decomposition model of an imperfect resonator is shown in Fig. 6.20. The elliptical deformation $A$ of the resonator is expressed as follows:

$$A(\theta) = A_0 \cos(2\theta) \tag{6.18}$$

where $A_0$ represents the maximum deformation at $\theta = 0°$.

In an imperfect resonator, its vibratory axis is not fully consistent with the direction of the elliptical deformation axis. Let $M$ and $N$ represent the vibration amplitudes

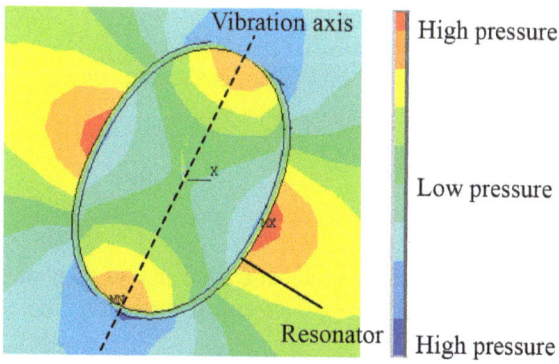

**Fig. 6.19** Acoustic pressure distribution during vibrations

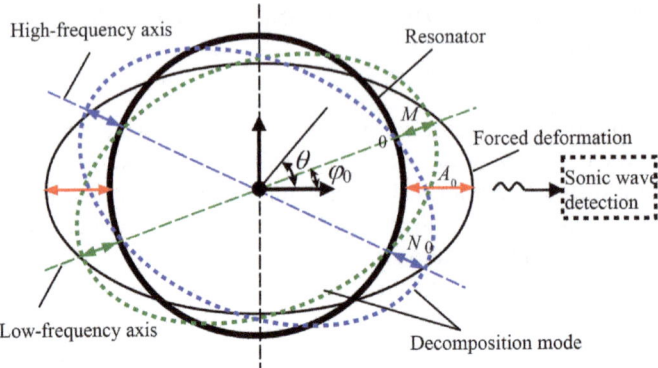

**Fig. 6.20** Vibration decomposition model of an imperfect resonator

along the high-frequency axis and low-frequency axis respectively. Under non-ideal conditions, the drive mode and sense mode are excited at the same time, and their mode shapes are as follows

$$M(\theta, t) = M_0 \cos 2(\theta - \varphi_0) \cos \omega_1 t$$
$$N(\theta, t) = N_0 \sin 2(\theta - \varphi_0) \cos \omega_2 t \qquad (6.19)$$

where $M_0$ and $N_0$ represent the maximum amplitudes; $\omega_1$ and $\omega_2$ are the respective natural frequencies. According to the principle of mode superposition, when the time variable $t$ is 0, $A_0$ can be expressed as

$$A_0 = M_0 \cos(2\varphi_0) + N_0 \sin(2\varphi_0) \qquad (6.20)$$

Energy loss is not considered in the short-term. Based on the principle of conservation of energy, the following can be derived:

$$\int_0^{2\pi} \left( A + \frac{\partial^2 A}{\partial \theta^2} \right)^2 d\theta = \int_0^{2\pi} \left( M + \frac{\partial^2 M}{\partial \theta^2} \right)^2 d\theta + \int_0^{2\pi} \left( N + \frac{\partial^2 N}{\partial \theta^2} \right)^2 d\theta \qquad (6.21)$$

Equations (6.20) and (6.21) are solved simultaneously, with the solution shown as follows:

$$M_0 = A_0 \cos(2\varphi_0)$$
$$N_0 = A_0 \sin(2\varphi_0) \qquad (6.22)$$

If the damping coefficient $\xi$ is taken into account, too, the radial displacement $w$ can be evaluated using the mode superposition method. This variable is expressed as:

$$w = A_0 e^{-\psi \xi t} \sqrt{\cos^2(2\theta) + \sin(4\varphi_0)\sin 4(\theta - \varphi_0)\cos^2\left(\frac{\Delta \omega t}{2}\right)} \cdot \sin(\omega_1 t + \Theta)$$

$$(6.23)$$

where $\xi = 1/2Q$, $\Psi$ is a coefficient related to the radius $R$, the density $\rho$, and the flexural stiffness $K$ of the resonant ring, $\psi = \sqrt{\frac{K}{\rho R^4}}$, $\Delta \omega = \omega_2 - \omega_1$. The vibration phase angle is as follows:

$$\Theta = \arctan\left[\frac{\sin(2\varphi_0)\sin 2(\theta - \varphi_0)\sin \Delta \omega t}{\cos(2\varphi_0)\cos 2(\theta - \varphi_0) + \sin(2\varphi_0)\sin 2(\theta - \varphi_0)\cos \Delta \omega t}\right] \quad (6.24)$$

Equations (6.23) and (6.24) contain frequency split $\Delta \omega$, modal offset $\varphi_0$ and quality factor. Because acoustic pressure $P$ is directly proportional to the vibration amplitude and the velocity of the resonator, the acoustic pressure is evaluated as follows:

$$P = k\frac{dw}{dt} \quad (6.25)$$

where $k$ represents the coupling coefficient between the resonator vibrations and the acoustic pressure. Therefore, the parameters of the resonator can be calculated inversely by using Eqs. (6.23) and (6.24).

The resonator parameters are detected according to the abovementioned mode detection principle. The acoustic test method contains the following steps:

Step 1: Drive the resonator to vibrate freely. Fix the resonator on a platform and excite its edge to generate initial deformation and make it vibrate freely. Record the peening location to facilitate the identification of the standing wave position.

Step 2: Collect acoustic signals using an acoustic pressure sensor. For ease of measurement, keep the center of the resonator, the peening location, and the acoustic pressure sensor on the same straight line. Try to place the acoustic pressure sensor as close to the edge of the resonator as possible in order to reduce acoustic losses in the air.

Step 3: Signal processing. Read the collected acoustic signals, and use the least square method to fit the acoustic signals to calculate the resonator's characteristic modal parameters.

The acoustic signals collected by the resonator are shown in Fig. 6.21a. As the figure shows, a strain of continuously evanescent oscillatory waves is formed during the free vibration of the resonator. The strain of waves is synthesized from the vibrations in both drive mode and sense mode. These acoustic signals are greatly deformed within the first 0.5 s, and this is mainly due to the great disturbances caused by the initial shock vibration. However, this type of non-ideal signal attenuates quickly, so

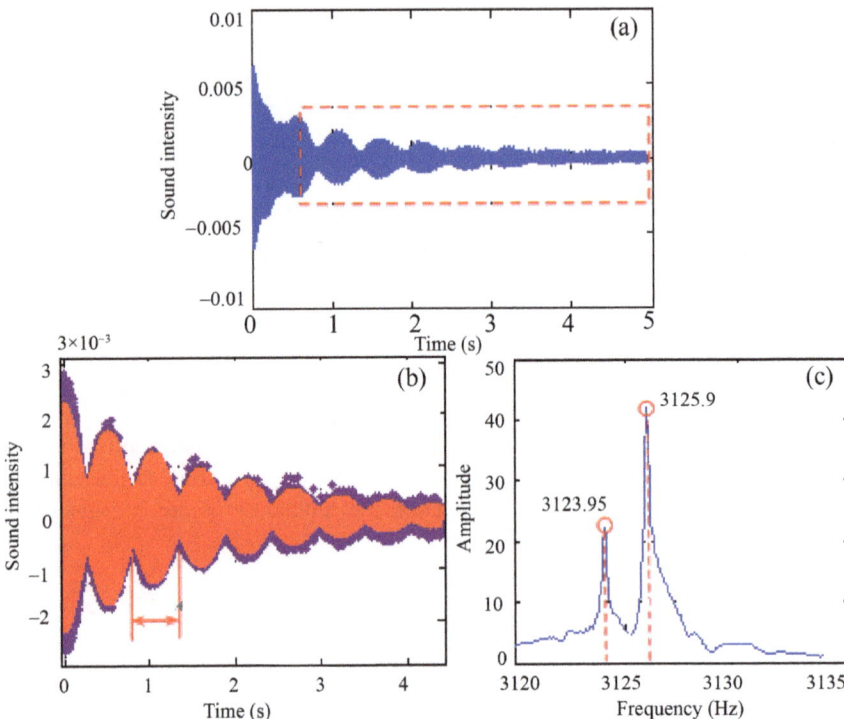

**Fig. 6.21** Detection and analysis of acoustic wave signals **a** acoustic signals; **b** fitted signals; **c** frequency spectrum analysis

in order to improve the precision of the fit, only signals between 0.5 and 5 s are selected for this test.

A parameter analysis is then performed in mathematical software. The least-square method is used to perform parameter identification. The results of the fit are shown in Fig. 6.21b. The results of the parameter fit shows that the frequency split of the resonator is 1.93 Hz, the quality factor is 3499.4, and the modal drift is 13.5°. The accuracy of parameter identification is closely related to the precision of the collected acoustic signals. It is also found that as the sampling time is extended, the signal–noise ratio is decreased. If filtering is performed, the identification accuracy can be further improved.

The signal spectrum can be analyzed by Fourier transform. The results are shown in Fig. 6.21c. It can be seen that the signal frequencies of the two trains of standing waves are 3231.95 Hz and 325.9 Hz respectively. The frequency split of the resonator is thus 1.95 Hz, and this is basically identical to the results of the parameter fitting. In addition, the spectrum amplitude shown in Fig. 6.21c also represents the magnitude of the vibrational energy.

Compared with traditional detection methods, acoustic detection has relatively high accuracy. Moreover, it can be easily performed, as long as a low-cost MEMS microphone acoustic pressure sensor is used. The cost of the detection tests can be effectively reduced, while detection efficiency can be improved.

# References

1. Rourke, A. K., McWilliam, S., & Fox, C. H. J. (2001). Multi-mode trimming of imperfect rings. *Journal of Sound and Vibration, 248*(4), 695–724.
2. Xi, X., Wu, Y. L., Wu, X. M., Tao, Y., & Wu, X. Z. (2012). Investigation on standing wave vibration of the imperfect resonant shell for cylindrical gyro. *Sensors and Actuators A: Physical, 179*, 70–77.
3. Jianqiu, W. (2014). *Research on non-contact driving and detection technology of CVGs.* National University of Defense Technology.
4. Bingjie, Z. (2011). *Analysis of the vibration characteristics of cup-shaped gyroscope resonators and research on trimming techniques.* National University of Defense Technology.
5. Xi, X., Wu, Y., Zhang, Y., Wu, X., Zheng, Y., Wu, X. (2014). A simple acoustic method for modal parameter measurement of the resonator for vibratory shell gyroscope. *IEEE Sensors Journal, 14*(11), 9.

# Chapter 7
# Closed Loop Control of CVGs

The measurement and control circuit is the key to angular velocity measurement within the CVG, and its performance directly influences the gyroscope's ability to detect angular velocity. This chapter introduces the measurement and control technologies based on the structure of the gyroscope. This includes model identification, drive control, balance control and angular velocity detection.

## 7.1 Operating Principles of the Measurement and Control Circuit

The high-precision resonator is the most important component of a gyroscope, though peripheral unit integration is also critical. The main functions of the measurement and control circuit include (1) exciting the drive mode of the resonator so that it can vibrate in the form of stable standing waves; (2) suppressing vibrations in sensing mode so that the gyroscope can have a better responsive bandwidth; (3) demodulating angular velocity from force in order to rebalance signals and provide an output signal for angular velocity.

Figure 7.1 shows the fundamental circuit principle of the CVG. This consists of two main sections: the drive loop and the force-to-balance loop. The closed drive loop of the resonator is primarily used for frequency tracking and amplitude control, while the force-to-balance loop is used for sensing mode suppression and angular velocity demodulation.

Near the natural frequency $\omega_0$, the vibration displacement $w$ of the resonator should be

$$w = A \cos(\omega_d t - \phi) \cos(2\theta) \tag{7.1}$$

© National Defense Industry Press 2021

X. Wu et al., *Cylindrical Vibratory Gyroscope*, Springer Tracts in Mechanical Engineering, https://doi.org/10.1007/978-981-16-2726-2_7

**Fig. 7.1** Circuit principle

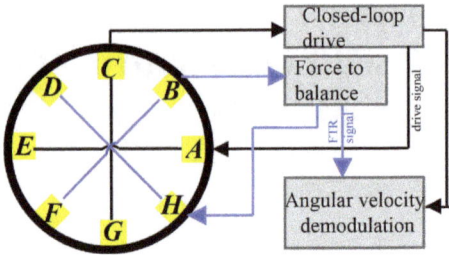

where $A$ represents the amplitude of the drive signal; $\omega_d$ represents the frequency of the drive signal; $\phi$ represents the phase difference between drive and response. This can be calculated by means of the following equation:

$$\phi = a\tan\frac{2\xi\gamma_1}{1 - \gamma_1^2} \tag{7.2}$$

where $\gamma_1 = \omega_0/\omega_d$. It can be seen that, if the tracks of the natural frequency of the resonator need to be followed, we can let $\gamma_1 = 1$, meaning the phase difference $\phi$ between drive and detection must be equal to 90°.

A phase shifter can be used to process the drive signal phase. In reality, the actual phase shift $\phi$ of the resonator will be:

$$\phi = \phi_d + \phi_{delay} \tag{7.3}$$

where $\phi_{delay}$ represents a fixed phase delay caused by the adhesion between various filters, as well as between the drive/detection electrodes and the resonator. Therefore, the parameter $\phi_d$ of the phase shifter can be actively adjusted so that phase shift can be kept at an ideal level of around 90°.

The amplitude of the resonator is controlled by a PID. The amplitude of the resonator is directly proportional to the voltage of the piezoelectric electrodes. As a result of this, the detected electric signal difference between the reference voltage and the piezoelectric electrode voltage is chosen as an input to the PID controller so that the displacement of the resonator will remain constant. The overall control circuit diagram is shown in Fig. 7.2.

Angular velocity information can be extracted by demodulating the signal from the force-to-rebalance circuit and the drive mode signal from the gyroscope. The core component of the demodulation circuit is a multiplier, which is taken as having ideal multiplication characteristics and outstanding overload capacity. An output voltage corresponding to the angular velocity can be obtained after the detection signal generated by the Coriolis effect is demodulated with the multiplier and then put through a low-pass filter and amplifying circuit.

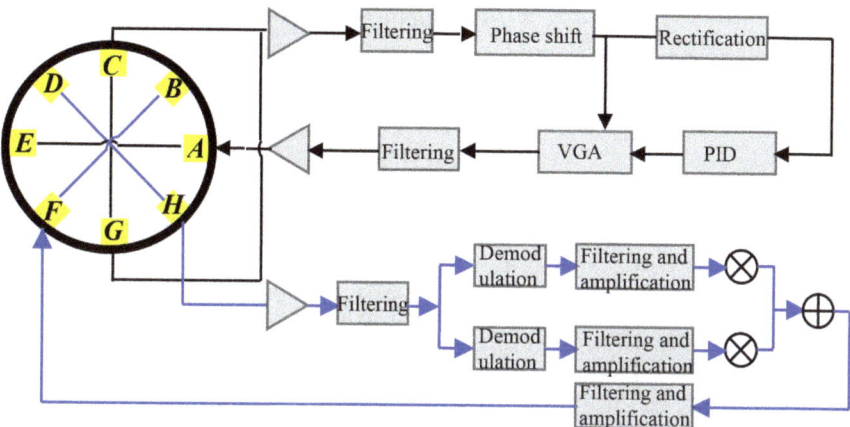

**Fig. 7.2**   Control circuit block diagram

## 7.2   Equivalent Circuit Models of Resonators

### 7.2.1   Equivalent Circuit Modeling for Resonators

There will be an output voltage and an input voltage at the gyroscope head when the drive mode and sensitivity mode of the resonator are excited through both the piezoelectric effect and the inverse piezoelectric effect. Its electrical characteristics can be described using an equivalent circuit model for the resonator. This transforms its complex mechanical vibrations into electrical signals that can be implemented in circuit components. In this way, a transfer function can be established for the gyroscope to provide a basis for the design of a measurement and control circuit [1].

The vibration of the CVG resonator relies on the piezoelectric effect. There is not only an input voltage but also an output voltage on its piezoelectric electrodes. The resonator, which is in itself an elastomer, has an unlimited number of modes, and the dynamic characteristics of each mode, similar to that of the *RLC* circuit, are dynamic characteristics of a second-order system. Therefore, the characteristics of the CVG resonator response to the input voltage frequency can be described by means of an equivalent circuit model, as shown in Fig. 7.3.

In Fig. 7.3, every branch in the equivalent circuit model describes one vibration mode of the resonator. $R_i$, $C_i$ and $L_i$ are the equivalent resistance, equivalent capacitance and equivalent inductance of the resonator, as related to the mechanical properties of the resonator. They correspond to the $i$th modal mass and stiffness of the resonator, respectively. $R_i$ represents the equivalent resistance of the resonator. This is related to the mechanical loss of the material and corresponds to the $i$th modal damping. In addition, the shunt capacitance $C_0$ represents the static capacitance among the piezoelectric electrodes. Although the dynamic response of the resonator is composed of an infinite number of modal responses superimposed on

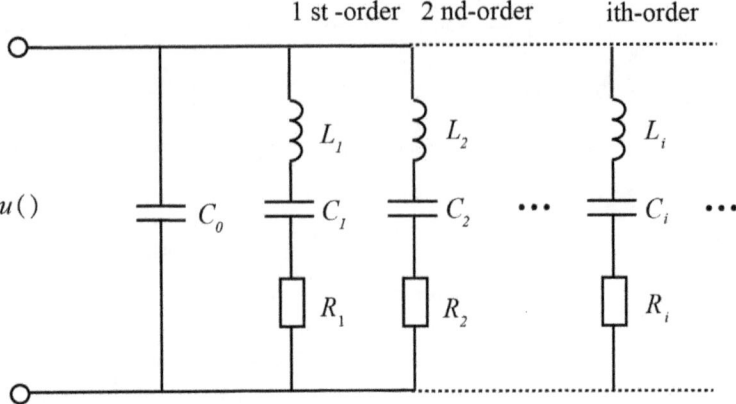

**Fig. 7.3**  Equivalent circuit model of the CVG resonator

one another, for the working modes of a CVG with a high quality factor, a highly
accurate dynamic response can be acquired by conducting a steady-state response
analysis of the driving force at the working mode frequency of the resonator. Therefore, the vibration of the resonator in its working modes excited by the voltage can be
simplified to an *RLC* branch. Figure 7.4 shows a simplified equivalent circuit model
of the CVG resonator.

According to analog circuit theory, the resonant frequency of the *RLC* branch in
Fig. 7.4 is:

$$\omega_i = \frac{1}{\sqrt{L_i C_i}} \tag{7.4}$$

The admittance of the *RLC* branch in Fig. 7.4 is:

$$Y_i(\omega) = \frac{1}{R_i + j\omega L_i + \frac{1}{j\omega C_i}}$$

**Fig. 7.4**  Simplified
equivalent circuit model of
the CVG resonator

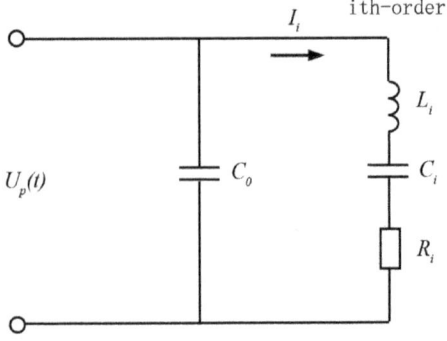

$$= \frac{\omega_p^2 R_i C_i^2 + j\left(\omega_p^3 R_i C_i^2 - R_i C_i\right)}{\left(1 - \omega_p^2 L_i C_i\right)^2 + \left(\omega_p R_i C_i\right)^2} = G_i(\omega) + j B_i(\omega) \tag{7.5}$$

where $G(\omega)$ represents conductance, while $B(\omega)$ represents susceptance.

To deduce the relationship between the dynamic model parameters and equivalent circuit parameters of the resonator, we can analyze the equivalent circuit characteristics of the CVG resonator from the perspective of energy conservation, that is to say based on the principle that the mechanical work done by the driving voltage of the gyroscope is equal to the electrical work. Let the driving voltage of the gyroscope $U_p(t) = U_{p0}\sin\omega_p t$, and the energy consumed by the $RLC$ branch in the equivalent circuit in a period of oscillation will be:

$$W_i = \int_0^{T_i} U_p(t) I_i(t) \mathrm{d}t = \int_0^{\frac{2\pi}{\omega}} U_p^2(t) \sin^2\left(\omega_p t\right) G_i(\omega) \mathrm{d}t$$

$$= \frac{\pi \omega_p R_i C_i^2 U_{p0}^2}{\left(1 - \omega_p^2 L_i C_i\right)^2 + \left(\omega_p R_i C_i\right)^2} = \frac{\pi \omega_p R_i C_i^2 U^2}{\left(1 - v_i^2\right)^2 + \left(\omega_p R_i C_i\right)^2} \tag{7.6}$$

The frequency ratio $v_i$ is:

$$v_i = \frac{\omega_p}{\omega_i} \tag{7.7}$$

According to the definition of the generalized coordinates and generalized force of the CVG resonator, the periodic work done by the generalized force in the drive mode of the resonator, i.e., in the $i$th mode, is equal to:

$$W_d = \int_0^{\frac{2\pi}{\omega_p}} F^* \cdot \dot{w}_A \mathrm{d}t$$

$$= \int_0^{\frac{2\pi}{\omega_p}} \frac{M_{p0}}{H} \sin\omega_p t \cdot w_{A\_st} \eta_d \omega_p \cos\left(\omega_p t - \beta_d\right) \mathrm{d}t$$

$$= \frac{\pi M_{p0}}{H} w_{A\_st} \eta_d \sin\beta_d \tag{7.8}$$

Based on this, the following can be derived:

$$W_d = \left(\frac{U_{p0}\Theta_{UM}}{H}\right)^2 \frac{\pi}{k_d^*} \frac{2\xi v_d}{(1 - v_d^2)^2 + 4\xi^2 v_d^2} \tag{7.9}$$

According to the principle of conservation of energy, Eq. (7.6) is equal to Eq. (7.9), and a relational expression can be derived:

$$
\begin{cases}
C_i = \left(\dfrac{\Theta_{UM}}{H}\right)^2 \dfrac{1}{k_d^*} \\
\omega_p R_i C_i = 2\xi\,v_d \\
\omega_i = \dfrac{1}{\sqrt{L_i C_i}} = \omega_d = \sqrt{\dfrac{k_d^*}{m_d^*}}
\end{cases}
\tag{7.10}
$$

According to the solution of Eq. (7.10), the relationship between the equivalent circuit parameters of the resonator and its physical parameters can be revealed as follows:

$$
\begin{cases}
U_p = F^*\left(\dfrac{H}{\Theta_{UM}}\right) = \dfrac{F^*}{n_{M-E}} \\
L_i = \left(\dfrac{H}{\Theta_{UM}}\right)^2 m_d^* = \dfrac{1}{n_{M-E}^2 m_d^*} \\
C_i = \left(\dfrac{\Theta_{UM}}{H}\right)^2 \dfrac{1}{k_d^*} = \dfrac{n_{M-E}^2}{k_d^*} \\
R_i = 2\xi\omega_d m_d^*\left(\dfrac{H}{\Theta_{UM}}\right)^2 = \dfrac{2\xi\omega_d m_d^*}{n_{M-E}^2}
\end{cases}
\tag{7.11}
$$

where $n_{M-E} = \dfrac{\Theta_{UM}}{H}$ represents the coefficient of conversion of the CVG resonator from the mechanical system to the electrical system.

As shown in Fig. 7.4, let the *RLC* branch current be $I_i(t)$, so the relationship between the circuit voltage and the current will be as follows:

$$
\begin{aligned}
U_p(t) &= L_i\frac{dI_i(t)}{dt} + R_i I_i(t) + \frac{1}{C_i}\int I_i(t)dt \\
&= L_i\frac{d^2 Q_i(t)}{dt^2} + R_i\frac{dQ_i(t)}{dt} + \frac{1}{C_i}Q_i(t)
\end{aligned}
\tag{7.12}
$$

According to Eqs. (7.11) and (7.12), the steady-state charge of the *RLC* branch under the action of the sinusoidal driving voltage can be evaluated as follows:

$$
Q_i(t) = U_{p0}C_i\eta_i \sin(\omega_p t - \beta_i)
\tag{7.13}
$$

where $\eta_i$ represents the magnification coefficient of the dynamic charges in the equivalent circuit; $\beta_i$ represents the lag phase response angle of the dynamic response in drive mode. These are expressed as follows:

$$
\begin{cases}
\eta_i = \dfrac{1}{\sqrt{\left(1-v_i^2\right)^2+\left(\omega_p R_i C_i\right)^2}} \\
\beta_i = \arccos\left(\dfrac{1-v_i^2}{\sqrt{\left(1-v_i^2\right)^2+\left(\omega_p R_i C_i\right)^2}}\right)
\end{cases}
\tag{7.14}
$$

The *RLC* branch current in the equivalent circuit can be equivalent to the generalized velocity of the resonator. According to Eq. (7.11), the relationship between the charges (currents) in the RLC branch and the generalized displacement (velocity) of the resonator can be determined as follows:

$$\begin{cases} Q_i = n_{M-E} w_A \\ I_i = n_{M-E} \dot{w}_A \end{cases} \tag{7.15}$$

In conclusion, the conversion relationship between the mechanical parameters and the electrical parameters of the CVG resonator is shown in Table 7.1.

To express the relationship between the driving voltage of the resonator in drive mode and the vibration detection signal in drive mode, we can introduce a current-controlled current source into the equivalent circuit of the resonator for transformation, as shown in Fig. 7.5.

The detection signal in the drive mode of the resonator shown in Fig. 7.5 is as follows:

$$U_{ds}(t) = \frac{1}{C_0} \int n_I I_i(t) dt = w_{A\_d} \frac{e_{31} h_p (h_b + h_p)}{2 H \varepsilon_{33} (R - R_0)} \sin(\omega_p t - \beta_d) \tag{7.16}$$

**Table 7.1** Comparison of equivalent mechanical–electrical parameters of the resonator

|  | Force-voltage | Displacement-electric quantity | Velocity-current | Mass-inductance | Stiffness-capacitance | Damping-resistance |
|---|---|---|---|---|---|---|
| Mechanical domain | $F^*$ | $w_A$ | $\dot{w}_A$ | $m_d^*$ | $k_d^*$ | $\xi$ |
| Electrical domain | $U_p$ | $Q_i$ | $I_i$ | $L_i$ | $1/C_i$ | $R_i$ |

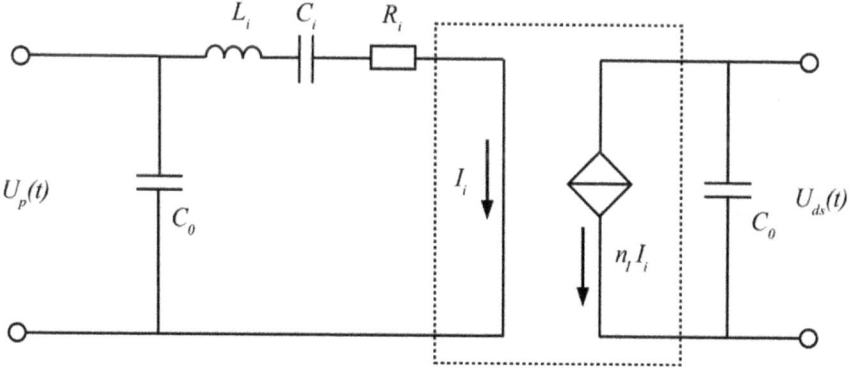

**Fig. 7.5** Equivalent circuit model of the gyroscope resonator for detection of drive mode

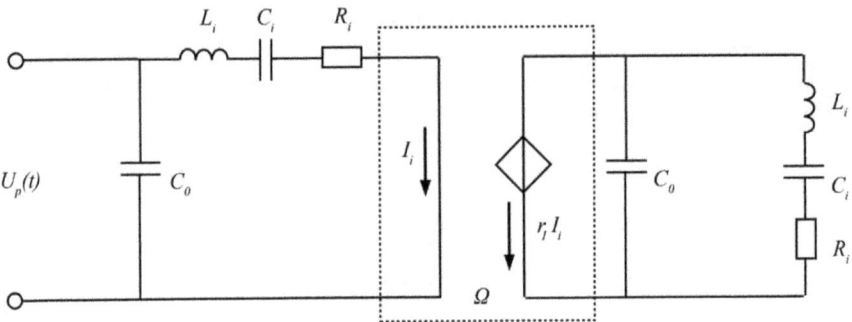

**Fig. 7.6** Coriolis force coupling and equivalent circuit model for CVG resonator

According to the solution of Eq. (7.16), the transfer current ratio $n_I$ of the current-controlled current source is as follows:

$$n_I = \frac{e_{31}b_p l_p (h_b + h_p)}{2\Theta_{UM}(R - R_0)} = \frac{c_{33}e_{31}}{c_{13}e_{33} - e_{31}c_{33}} \tag{7.17}$$

Since the drive mode and sensitivity mode of an ideal CVG resonator have the same dynamic characteristics, the sensitivity mode of the resonator can also be equivalent to the *RLC* circuit. What is more, the parameters are the same as the equivalent circuit parameters in the drive mode. To express the relationship between the vibration velocity and the Coriolis force on the resonator in drive mode, a current-controlled voltage source can be introduced into the equivalent circuit, as shown in Fig. 7.6.

The equivalent driving voltage of the resonator in sensitivity mode as shown in Fig. 7.6 can be calculated using the following equation:

$$U_c(t) = r_I I_i(t) = \frac{M_c}{\Theta_{UM}} = \frac{\Omega \dot{w}_A \Theta_{\Omega M}}{\Theta_{UM}} = \frac{\Omega I_i \Theta_{\Omega M}}{n_{M-E}\Theta_{UM}} \tag{7.18}$$

Using Eq. (7.18), the transfer resistance $r_I$ of the current-controlled voltage source can be evaluated as follows:

$$r_I = \Omega \frac{\Theta_{\Omega M}}{n_{M-E}\Theta_{UM}} = \Omega r_\Omega \tag{7.19}$$

Like the detection signal in drive mode, the signal in sensitivity mode can also be transformed by introducing a current-controlled voltage source into the equivalent circuit of the resonator. The specific model is similar to the equivalent model of the detection signal in drive mode. Figure 7.7 shows the equivalent drive and detection circuits of the CVG resonator.

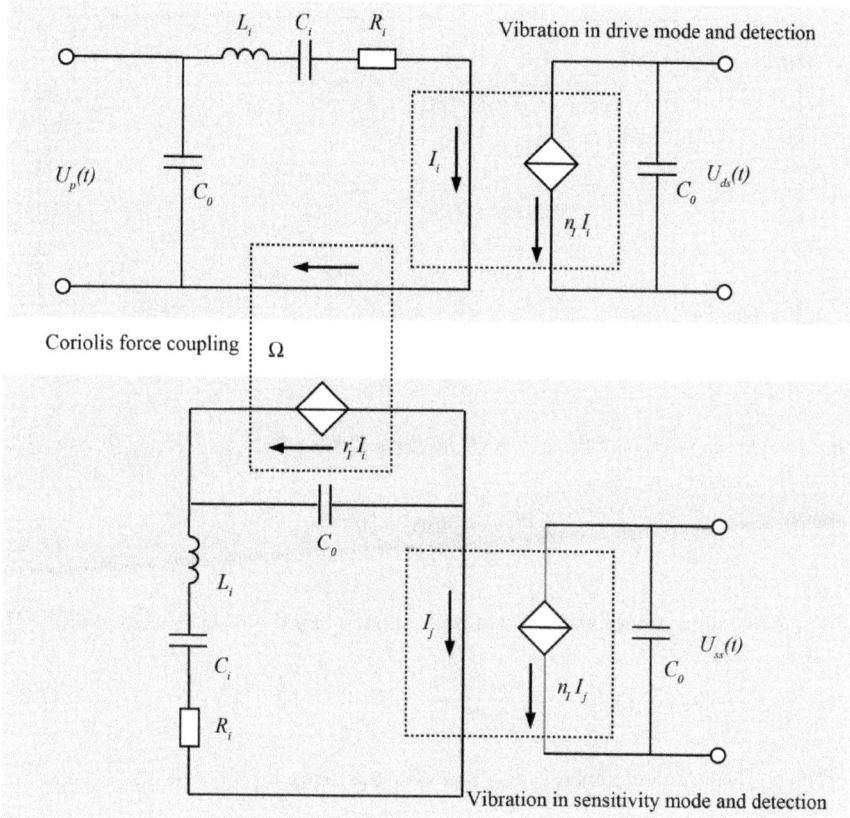

**Fig. 7.7** Equivalent drive and detection circuits of the CVG resonator

## 7.2.2 Equivalent Parameter Identification for Resonators

According to the foregoing process of equivalent circuit modeling, the equivalent circuit parameters of the resonator can be found as long as the frequency–response curve near the drive mode of the CVG resonator is known. The vertical coordinate of the frequency–response curve in drive mode is transformed into an output voltage value as shown in Fig. 7.8.

The quality factor of the system can be determined by locating the half-power point of the system on the frequency–response curve in Fig. 7.8.

$$Q = \frac{1}{2\xi} = \frac{\omega_i}{\omega_2 - \omega_1} = \frac{3953.98}{3954.21 - 3953.77} = 8786 \qquad (7.20)$$

By reference to the technical parameters of the piezoelectric electrodes of the resonator, the transfer current ratio $n_I$ between the static capacitance of the resonator and the equivalent current-controlled current source can be evaluated as follows:

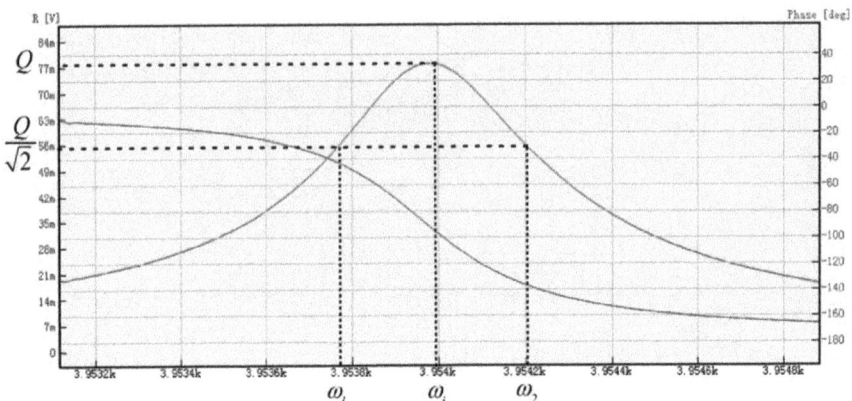

**Fig. 7.8** Frequency–response curve for CVG resonator in drive mode

$$\begin{cases} C_0 = 1.486 \times 10^{-9}\,\text{F} \\ n_I = 0.272 \end{cases} \tag{7.21}$$

The equivalent capacitance $C_i$ can be evaluated based on Eqs. (7.10) and (7.13):

$$C_i = \frac{U_{ds-\text{max}} C_0 n_I}{U_{p0} Q} = 9.707 \times 10^{-14}\,\text{F} \tag{7.22}$$

The equivalent inductance $L_i$ can be evaluated using Eq. (7.10):

$$L_i = \frac{1}{\omega_i^2 C_i} = 1.669 \times 10^4\,\text{H} \tag{7.23}$$

The equivalent resistance $L_i$ can be evaluated using Eq. (7.10):

$$R_i = \frac{1}{Q\omega_i C_i} = 4.721 \times 10^4\,\Omega \tag{7.24}$$

According to the equivalent circuit model of the resonator, a transfer function can be established for the resonator from its driving voltage to the detection voltage, as follows:

$$\begin{aligned} H_d(s) &= \frac{U_{ds}(s)}{U_p(s)} = \frac{1}{R_i + sL_i + \frac{1}{sC_i}} \cdot \frac{n_I}{sC_0} \\ &= \frac{1}{L_i C_i s^2 + R_i C_i s + 1} \cdot \frac{n_I C_i}{C_0} \end{aligned} \tag{7.25}$$

The equivalent circuit parameters obtained from Eqs. (7.20)–(7.24) are substituted into Eq. (7.25), deriving a transfer function for the resonator from its driving voltage to the detection voltage, as follows:

$$H_d(s) = \frac{1.777 \times 10^{-5}}{1.620 \times 10^{-9}s^2 + 4.583 \times 10^{-9}s + 1} \tag{7.26}$$

The equivalent circuit parameters of the resonator are identified according to the frequency–response curve, deriving a transfer function for the resonator in its drive mode.

## 7.3 Drive Control Design

To ensure the high sensitivity and stability of the gyroscope, the vibration amplitude of the gyroscope in drive mode must be increased as much as possible and kept constant when the device is in drive control mode. The present section introduces the drive control technology of the CVG, including frequency control and amplitude control.

### 7.3.1 Resonant Excitation in Resonator Drive Mode

The driving voltage frequency of the resonator has a great influence on vibration amplitude when the device is in drive mode. Only when the driving voltage frequency is equal to the natural frequency of the resonator in its operating modes can a maximum vibration amplitude be obtained. This trait is especially noticeable when a larger quality factor is present. However, when the gyroscope is in operation, changes in the external temperature or material characteristics can affect the natural frequency of the resonator. As a result, during drive control of the gyroscope, the frequency of the driving voltage should always be able to track the drive mode frequency of the resonator.

According to the phase-frequency characteristics listed in the transfer function of the resonator, no matter how the drive mode frequency $\omega_p$ changes, the phase shift will always be equal to $-90°$ at the resonant frequency. Therefore, the present section introduces a self-excitation method based on typical phase control, which enables the transfer function of the resonator in drive mode to maintain a phase shift of $-90°$ to the input signal, ensuring that the driving voltage frequency $\omega_p$ always tracks the resonant frequency $\omega_d$.

Figure 7.9 shows a self-excited signal flow diagram based on phase control. This is composed of an equivalent circuit module of the resonator, an amplifier and a phase shifter; when the loop meets the condition of self-excited oscillations, the resonator resonates.

The amplitude equilibrium condition and phase equilibrium condition must be met in order that self-excited oscillations can be generated in the loop. First of all, the amplitude equilibrium conditions for self-excited oscillations within the loop are analyzed.

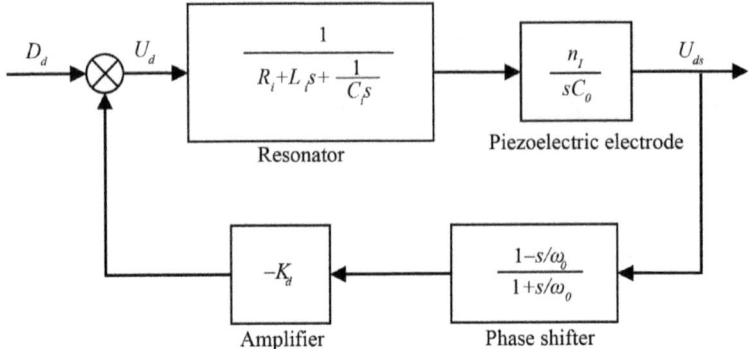

**Fig. 7.9** Phase control-based self-excited signal flow diagram

According to Eq. (7.25), the gain and phase of the equivalent circuit module of the resonator are as follows:

$$\begin{cases} G_i(\omega) = |H_d(j\omega)| = \dfrac{1}{\sqrt{(1-v_i^2)^2+\left(\frac{v_i}{Q_i}\right)^2}} \cdot \dfrac{n_I C_i}{C_0} \\[4mm] \varphi_i(\omega) = -\arccos\dfrac{1-v_i^2}{\sqrt{(1-v_i^2)^2+\left(\frac{v_i}{Q_i}\right)^2}} \end{cases} \tag{7.27}$$

where

$$Q_i = \frac{1}{R_i}\sqrt{\frac{L_i}{C_i}}, \quad v_i = \frac{\omega}{\omega_i}, \quad \omega_i = \frac{1}{\sqrt{L_i C_i}} \tag{7.28}$$

The gain and phase of the phase shifter are as follows:

$$\begin{cases} G_s(\omega) = \left|\dfrac{1-j\omega/\omega_0}{1+j\omega/\omega_0}\right| = 1 \\[4mm] \varphi_s(\omega) = -\arccos\dfrac{1-(\omega/\omega_0)^2}{\sqrt{(1-(\omega/\omega_0)^2)^2+(2\omega/\omega_0)^2}} \end{cases} \tag{7.29}$$

In order that self-excited oscillations can be generated in the loop, the entire drive loop needs to meet the amplitude equilibrium condition, i.e.,

$$G_{loop} = G_i(\omega) \cdot G_s(\omega) \cdot K_d \geq 1 \tag{7.30}$$

According to Eqs. (7.27) and (7.29), given any $\omega$, the gain of the phase shifter $G_s(\omega) \equiv 1$. If the $Q_i$ value of the resonator is large, then the equivalent circuit module gain $G_i(\omega) \geq \frac{n_I C_i}{C_0}$, for any given $\omega$. Therefore, in order that the loop gain $G_{loop} \geq 1$ for any $\omega$, we will obtain:

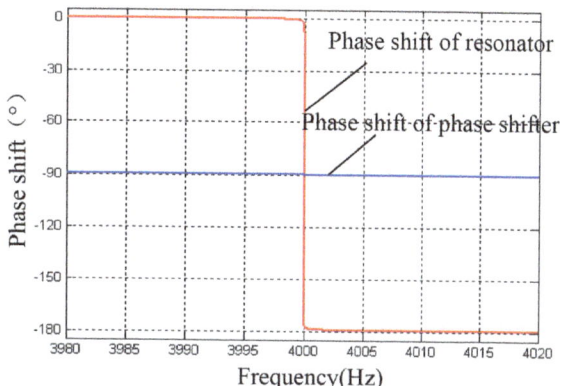

**Fig. 7.10** Equivalent circuit model of the resonator with phase shift of the phase shifter

$$K_d \geq \frac{C_0}{n_I C_i} \tag{7.31}$$

This is the amplitude equilibrium condition for the generation of self-excited oscillations in the drive control loop of the resonator.

Considering that the $Q_i$ value of the gyroscope resonator is generally very high, the phase shift of the transfer function of the resonator's equivalent circuit module changes tremendously as it gets near to $\omega_i$. Similarly, the phase shift of the phase shifter varies among signals of different frequencies, so that the phase equilibrium condition can be considered to be within a small range near the resonant frequency $\omega_i$. Let $Q_i = 5000$. When $\omega_i = \omega_0 = 4000$ Hz, the equivalent circuit model of the resonator and phase shift of the phase shifter are drawn in a range of $0.995 < \omega/\omega_0 < 1.005$, as shown in Fig. 7.10.

As can be seen in Fig. 7.10, within this frequency range, the phase shift of the drive shaft's transfer function varies greatly, while the phase shift of the phase shifter is approximately constant. In order that self-excited oscillations can be generated in the loop, the entire drive loop needs to meet the phase equilibrium condition, i.e.,

$$\varphi_{loop} = \varphi_i(\omega) + \varphi_s(\omega) - \pi = 2n\pi \tag{7.32}$$

Near $\omega_i$, the phase shift of the phase shifter has a value range of $-\pi$, 0. The phase shift of the transfer function of the resonator's equivalent circuit module has a value range of $-\pi$, 0. Using Eq. (7.32), the following can be derived:

$$\varphi_i(\omega) = (2n+1)\pi - \varphi_s(\omega) \tag{7.33}$$

A tangent value is assigned to each side of Eq. (7.33), meaning

$$\left(\frac{\omega}{\omega_d}\right)^2 + \frac{1}{Q_d \tan \varphi_s} \frac{\omega}{\omega_d} - 1 = 0 \tag{7.34}$$

The oscillation frequency $\omega_r$ of the drive loop can be evaluated by solving Eq. (7.34):

$$\frac{\omega_r}{\omega_i} = \frac{-1}{2Q_i \tan \varphi_s} + \sqrt{1 + \frac{1}{4Q_i^2 (\tan \varphi_s)^2}} \tag{7.35}$$

As can be seen from Eq. (7.35), the oscillation frequency $\omega_r$ of the loop is related to the quality factor $Q_i$ of the resonator and the phase shift $\varphi_s$ of the phase shifter. The oscillation frequency of the loop can be changed by adjusting the characteristic frequency of the phase shifter. This is the basic principle of phase control technology.

By comparing the loop gain corresponding to the resonant frequency $\omega_r$ of the loop evaluated in Eq. (7.35) with the loop gain corresponding to the resonant frequency $\omega_i$ of the resonator in its drive mode, the loop gain ratio can be calculated as

$$\frac{|H_d(j\omega_r)|}{|H_d(j\omega_i)|} = \frac{1}{Q_i \sqrt{\left(1 - \left(\frac{\omega_r}{\omega_i}\right)^2\right)^2 + \left(\frac{1}{Q_i}\frac{\omega_r}{\omega_i}\right)^2}} \tag{7.36}$$

For a gyroscope resonator with a quality factor of $Q_i = 8786$ and a resonant frequency in drive mode of $\omega_i = 3954$ Hz, the relationship between the oscillation frequency $\omega_r$ of its loop, the phase shift $\varphi_s$ of the phase shifter and the loop gain ratio is shown in Fig. 7.11.

As shown in Fig. 7.11, when the phase shift of the phase shifter changes from $-100°$ to $-80°$, the maximum difference between the oscillation frequency of the loop and the frequency of the resonator in drive mode is only about 0.14 Hz. At this time, the loop gain is around 94.4% of the loop gain corresponding to the resonant frequency $\omega_i$ of the resonator in drive mode, indicating that the drive loop is effectively in its resonant state.

In summary, phase control technology-based self-excitation enables the oscillation frequency of the drive loop to remain very close to the resonant frequency of the

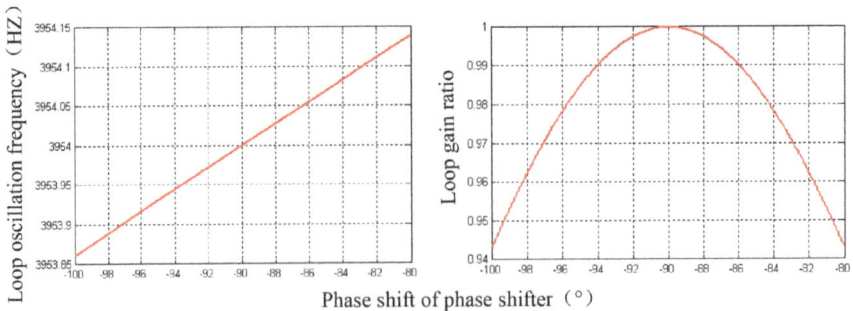

**Fig. 7.11** Relationship between the phase shift of the phase shifter, the oscillation frequency of the loop and the loop gain ratio

resonator in drive mode, thus ensuring that the drive loop stays in its resonant state. Moreover, the disturbance from the phase shifter has a very minor effect on the gain of the drive loop, thereby ensuring the frequency stability of the drive loop.

### 7.3.2 Amplitude Stabilization Control in Resonator Drive Modes

The preceding section introduced the phase control-based phase equilibrium and amplitude equilibrium conditions for the self-excitation drive loop of the resonator. When the gain of the entire loop is greater than 1, the signal amplitude of the loop will increase quickly after resonance is established and finally reach saturation. Loop gain control therefore needs to be enhanced so that gain will be maintained at 1, ensuring stable oscillation of the loop. Furthermore, when the gyroscope is used, the quality factor and resonance frequency of its resonator will change with temperature, thermal stress and damping. As a result, the vibration amplitude of the drive shaft will change accordingly. This means that a vibration amplitude control unit should be designed so that the amplitude of the resonator in drive mode can be kept constant, enabling the gyroscope to eventually achieve good bias stability and linearity.

According to the principle of operation of the CVG, the vibration amplitude of the resonator can be detected from the output voltage of the piezoelectric electrode for detection in drive mode. Therefore, the output voltage of the piezoelectric detection electrode in drive mode is taken as a variable for closed-loop amplitude control. The closed-loop control-based signal flow diagram is shown in Fig. 7.12.

Figure 7.12 shows an automatic gain control loop (AGC) transformed from the resonant excitation loop with three new modules: a rectifier, a low-pass filter and a controller. These have been added on the premise that they will not change the phase characteristics of the resonance loop. The detection voltage $U_{ds}$ of the resonator in drive mode generates a vibration amplitude signal $U_a$ through the rectifier and low-pass filter. After this, the signal is superimposed on the reference signal $U_r$ (the control target of the resonator amplitude) as an input signal of the controller, while the output signal $U_{dc}$ of the controller is modulated with the resonant frequency signal $U_s$ to form the drive signal $U_d$ in resonator drive mode.

As it contains nonlinear elements such as a rectifier, the resonator amplitude control system is itself nonlinear. To simplify the design of the controller, the system is transformed into a linear system by means of signal amplitude extraction. This means that problems that cannot be tackled for a nonlinear system can also be solved.

When the drive loop is in a stable state, $\omega_r \approx \omega_i$. At this time, the phase shifter outputs a baseband signal $U_s \sin \omega_i t$. It is assumed that the controller outputs a step signal $U_{dc} = u(t)$, so that in drive mode, the resonator has a drive signal $U_d = u(t)U_s \sin \omega_i t$.

According to Eq. (7.25), a transfer function is established for the equivalent circuit module of the resonator in Fig. 7.12, as follows:

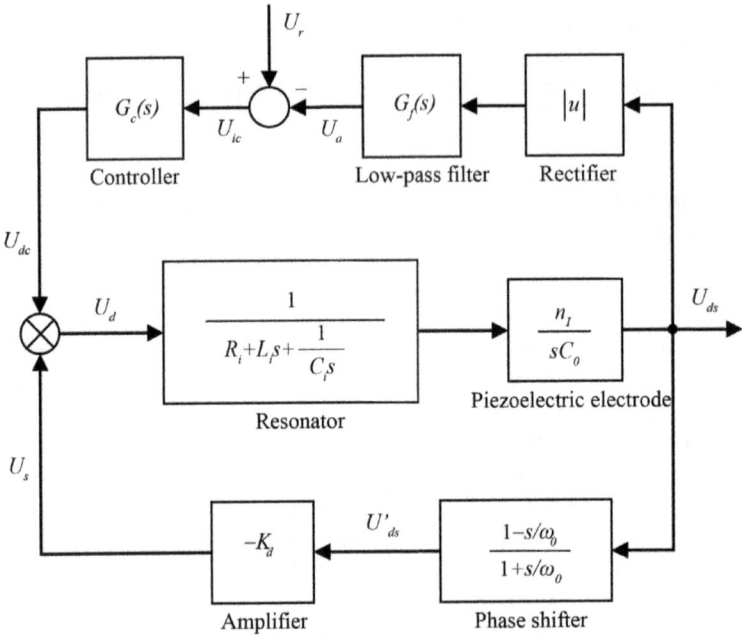

**Fig. 7.12** Closed-loop drive control-based signal flow diagram for the resonator

$$\frac{U_{ds}(s)}{U_d(s)} = \frac{1}{L_i C_i s^2 + R_i C_i s + 1} \cdot \frac{n_I C_i}{C_0} \tag{7.37}$$

Therefore,

$$\begin{aligned}
U_{ds}(s) &= \frac{1}{L_i C_i s^2 + R_i C_i s + 1} \cdot \frac{n_I C_i}{C_0} U_d(s) \\
&= \frac{1}{L_i C_i s^2 + R_i C_i s + 1} \cdot \frac{n_I C_i}{C_0} \frac{U_s \omega_i}{s^2 + \omega_i^2}
\end{aligned} \tag{7.38}$$

Equation (7.36) is an inverse-Laplace transformation, from which the following can be derived:

$$U_{ds}(t) \approx U_s \frac{n_I C_i Q_i}{C_0} \left( -\cos \omega_i t + e^{\left(-\frac{\omega_i t}{2 Q_i}\right)} \cosh\left( \frac{t\sqrt{(1 - 4 Q_i^2)\omega_i^2}}{2 Q_i} \right) \right) \tag{7.39}$$

The quality factor of the gyroscope resonator is much greater than 1, so the above equation can be simplified to:

$$U_{ds}(t) \approx U_s \frac{n_I C_i Q_i}{C_0} \left( -1 + e^{\left(-\frac{\omega_i t}{2 Q_i}\right)} \right) \cos \omega_i t \tag{7.40}$$

With the high-frequency components ignored after $U_{ds}(t)$ passes through the rectifier and low-pass filter, the amplitude signal $U_a(t)$ can be Laplace-transformed as follows:

$$U_a(s) = \frac{1}{s}\frac{2}{\pi}U_s\frac{n_I C_i Q_i}{C_0}\frac{\omega_i/2Q_i}{s + \omega_i/2Q_i}G_f(s) \tag{7.41}$$

For the controller $G_c(s)$, the transfer function of the entire controlled object is as follows:

$$G_d(s) = \frac{U_{ic}(s)}{U_{dc}(s)} = \frac{2}{\pi}U_s\frac{n_I C_i Q_i}{C_0}\frac{\omega_i/2Q_i}{s + \omega_i/2Q_i}G_f(s)$$
$$= K_e\frac{\omega_e}{s + \omega_e}G_f(s) \tag{7.42}$$

where $K_e$ represents the equivalent scale coefficient, and $\omega_e$ represents the equivalent frequency. Judging from Eq. (7.24), the controlled object $G_d(s)$ is an equivalent linear model.

The transfer function of the low-pass filter is defined as:

$$G_f(s) = \frac{\omega_f}{s + \omega_f} \tag{7.43}$$

The control signal flow diagram in Fig. 7.12 can be equivalent to the amplitude signal flow diagram in Fig. 7.13.

The closed-loop transfer function of the drive system shown in Fig. 7.13 is as follows:

$$T_d(s) = \frac{U_a(s)}{U_r(s)} = \frac{G_c(s)G_d(s)}{1 + G_c(s)G_d(s)} \tag{7.44}$$

A classic PID controller is used to control the amplitude of the CVG resonator in drive mode. A transfer function is established for the PID controller as follows:

$$G_c(s) = K_1 + K_2\frac{1}{s} + K_3 s \tag{7.45}$$

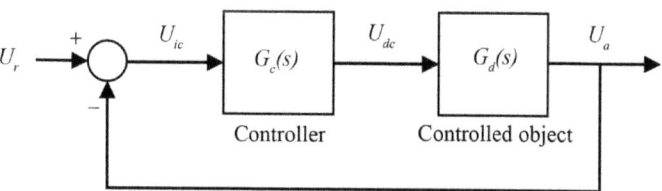

**Fig. 7.13**  Closed-loop control signal flow diagram of the resonator in drive mode

Equations (7.42), (7.43) and (7.45) are substituted into Eq. (7.44). The transfer function of the closed-loop system then takes the following form:

$$T_d(s) = \frac{\left(K_1 + K_2\frac{1}{s} + K_3 s\right)\left(K_e \frac{\omega_e}{s+\omega_e}\frac{\omega_f}{s+\omega_f}\right)}{1 + \left(K_1 + K_2\frac{1}{s} + K_3 s\right)\left(K_e \frac{\omega_e}{s+\omega_e}\frac{\omega_f}{s+\omega_f}\right)}$$

$$= \frac{K_3 K_e \omega_e \omega_f s^2 + K_1 K_e \omega_e \omega_f s + K_2 K_e \omega_e \omega_f}{s^3 + \left(\omega_e + \omega_f + K_3 K_e \omega_e \omega_f\right)s^2 + (1 + K_1 K_e)\omega_e \omega_f s + K_2 K_e \omega_e \omega_f}$$

(7.46)

The ITAE performance index is adopted for optimization of the controller design. The ITAE performance index refers to an integral over the product of time and absolute error. It is often applied to robust PID design. For a step response, the closed-loop transfer function is characterized by the optimal coefficient of a characteristic polynomial, as shown in Fig. 7.2.

The ITAE performance index-based PID controller design process is as follows: (1) Determine the natural frequency $\omega_0(T_s/\zeta\omega_0)$ of the closed-loop system based on the requirements of the adjustment time and damping ratio of the closed-loop system; (2) select an optimum closed-loop transfer function $T(s)$ according to the order of the closed-loop system and then determine 3 parameters of the PID controller; (3) select an appropriate prefilter so that there will be no zero points within the system.

As can be seen from Eq. (7.45), the drive and control system of the resonator is a third-order closed-loop system. According to the ITAE performance index-based third-order optimized characteristic polynomial shown in Table 7.2, the optimal transfer function of the closed-loop system is as follows:

$$T_d(s) = \frac{\omega_0^3}{s^3 + 1.75\omega_0 s^2 + 2.15\omega_0^2 s + \omega_0^3} \tag{7.47}$$

Equations (7.46) and (7.47) are then compared, with the following derived:

$$\begin{cases} \left(\omega_e + \omega_f + K_3 K_e \omega_e \omega_f\right) = 1.75\omega_0 \\ (1 + K_1 K_e)\omega_e \omega_f = 2.15\omega_0^2 \\ K_2 K_e \omega_e \omega_f = \omega_0^3 \end{cases} \tag{7.48}$$

**Table 7.2** Optimized characteristic polynomial based on ITAE

| System order | Optimized characteristic polynomial |
|---|---|
| First-order | $s + \omega_0$ |
| Second-order | $s^2 + 1.4\omega_0 s + \omega_0^2$ |
| Third-order | $s^3 + 1.75\omega_0 s^2 + 2.15\omega_0^2 s + \omega_0^3$ |
| Fourth-order | $s^4 + 2.1\omega_0 s^3 + 3.4\omega_0^2 s^2 + 2.7\omega_0^3 s + \omega_0^4$ |
| Fifth-order | $s^5 + 2.8\omega_0 s^4 + 5\omega_0^2 s^3 + 5.5\omega_0^3 s^2 + 3.4\omega_0^4 s + \omega_0^5$ |

$\omega_0$ is determined based on the start time of the CVG. The relationship between the natural frequency of the closed-loop system and the adjustment time and damping ratio of the system is as follows:

$$T_s = 4/\zeta\omega_0 \tag{7.49}$$

According to the transfer function of the resonator in drive mode, the PID parameters used for vibration amplitude control can be obtained. After this, the various parameters in the hardware circuit of the PID controller can be determined.

## 7.4 Detection and Control Methods

### 7.4.1 Force-To-Rebalance Techniques

According to the working principle of the CVG, after the angular velocity on the sensitive axis excites the drive mode of the resonator, the external angular velocity can be calculated by detecting the vibrations in sensitivity mode. However, the resonator's drive mode and sensitivity mode share the same frequency. In addition to this, since the quality factor of the resonator is very high, the dynamic performance of the gyroscope will suffer and nonlinearity will arise to a certain extent if angular velocity signals are acquired by the method adopted for other vibratory gyroscopes. In other words, the vibration of the resonator in sensitivity mode needs to be used directly as the output in order to calculate angular velocity. The present section introduces the force-to-balance circuit of the CVG. We will apply a control voltage to the piezoelectric electrode corresponding to the sensitivity mode of the resonator. This is down to suppress the vibration of the resonator in sensitivity mode and output force-to-balance control voltage signals. In this way, the external angular velocity can be calculated in order to improve the working bandwidth and linearity of the gyroscope [2, 3].

Because the natural frequency of the resonator in sensitivity mode is the same as the drive signal frequency of the resonator, and since the quality factor of the resonator is relatively high, the resonator is bound to vibrate significantly under the Coriolis force. Furthermore, the vibration phase is the same as the drive mode of the resonator, which will affect the stability of the drive mode. When a non-constant angular velocity signal is input into the gyroscope, the sensitivity mode of the resonator will remain stable for a long time, affecting the output of angular velocity signals from the gyroscope. The function of the force-to-balance circuit will offset the sensitivity mode of the resonator excited by the Coriolis moment by applying a control voltage to the resonator, thus generating a driving moment. That is to say, the output voltage of the piezoelectric electrode is equal to zero in sensitivity mode at any angular input velocity. Moreover, the resonator remains stable for an extremely short time, thus effectively increasing the working bandwidth of the

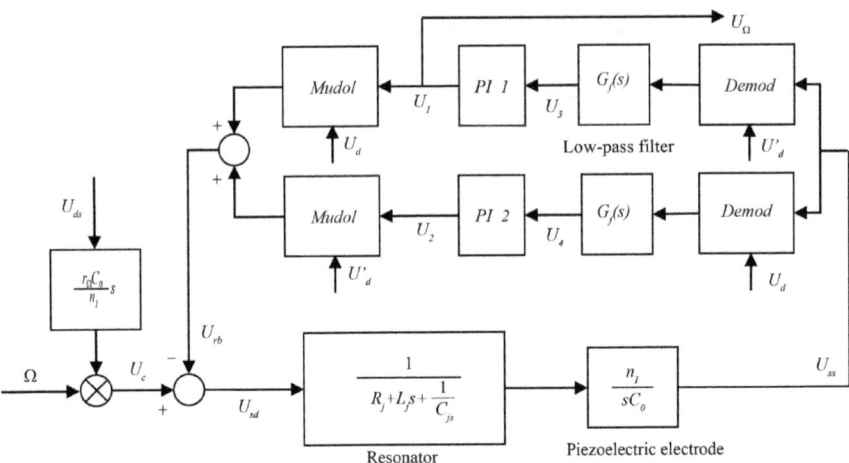

**Fig. 7.14** Signal flow diagram of the force-to-balance and angular velocity detection system of the resonator

resonator. Figure 7.14 is a signal flow diagram of the force-to-balance and angular velocity detection system of the resonator.

Based on the equivalent circuit model, the Coriolis force drives the resonator to vibrate in sensitivity mode, which can be equivalent to the driving voltage in sensitivity mode. Although the CVG basically resonates in sensitivity mode, any error in the measurement and control circuit of the gyroscope is bound to cause a small phase error between the output voltage of the resonator in drive mode and the theoretical value. The equivalent voltage of the Coriolis force is obtained based on Eqs. (7.16) and (7.18):

$$U_c(t) = \Omega I_i r_\Omega = \Omega \frac{r_\Omega C_0}{n_I} \frac{dU_{ds}(t)}{dt}$$
$$= U_{c1} \sin \omega_i t + U_{c2} \cos \omega_i t \tag{7.50}$$

Therefore, the effect of the voltage equivalent to the Coriolis force under non-ideal conditions should be taken into account when a force-to-balance circuit is designed, i.e. $U_{c1}\sin\omega_i t$ and $U_{c2}\cos\omega_i t$ should be force-balanced using two PI control branches. The entire force-to-balance circuit of the resonator, which also consists of nonlinear aspects such as demodulation and modulation, is a nonlinear system. To simplify the design of the controller, the system is transformed into a linear system by signal amplitude extraction.

Taking PI control loop 1 as an example, the controller output can be set to a step signal $U_1 = u(t)$, to create a drive signal $U_{sd} = u(t)U_d \sin \omega_i t$ for the resonator in sensitivity mode.

Similarly,

$$U_{ss}(t) \approx U_d \frac{n_1 C_i Q_i}{C_0} \left( -1 + e^{\left( -\frac{\omega_i t}{2Q_i} \right)} \right) \cos \omega_i t \tag{7.51}$$

The high-frequency components of $U_{ss}(t)$ are ignored after it is demodulated and low-pass filtered, before the input signal $U_3(t)$ of the PI controller is Laplace-transformed:

$$U_3(s) = U_d \frac{n_1 C_i Q_i}{C_0} \frac{1}{s} \frac{\omega_i/2Q_i}{s + \omega_i/2Q_i} G_f(s) \tag{7.52}$$

For the PI controller $G_c(s)$, the transfer function of the entire controlled object is as follows:

$$\begin{aligned}
G_s(s) &= \frac{U_3(s)}{U_1(s)} = U_s \frac{n_1 C_i Q_i}{C_0} \frac{\omega_i/2Q_i}{s + \omega_i/2Q_i} G_f(s) \\
&= K_s \frac{\omega_e}{s + \omega_e} G_f(s)
\end{aligned} \tag{7.53}$$

where $K_e$ represents the equivalent scale coefficient and $\omega_e$ represents the equivalent frequency. As can be seen, the controlled object $G_s(s)$ is an equivalent linear model.

The PI controller is used as the force-to-balance controller of the CVG. The transfer function of the two PI controllers can be set as follows:

$$\begin{cases}
G_{c1}(s) = K_1 + K_2 \dfrac{1}{s} \\
G_{c2}(s) = K_3 + K_4 \dfrac{1}{s}
\end{cases} \tag{7.54}$$

The transfer function of the low-pass filter is defined as:

$$G_f(s) = \frac{\omega_f}{s + \omega_f} \tag{7.55}$$

By reference to Eq. (7.47), the closed-loop transfer function of the system shown in Fig. 7.14 can be derived as follows:

$$T_s(s) = \frac{U_{rb}(s)}{U_c(s)} = \frac{G_{c1}(s)G_s(s)}{1 + G_{c1}(s)G_s(s)} \tag{7.56}$$

Equations (7.53), (7.54) and (7.55) are substituted into Eq. (7.44). The transfer function of the closed-loop system takes the following form:

$$T_s(s) = \frac{\left(K_1 + K_2\frac{1}{s}\right)\left(K_s \frac{\omega_e}{s+\omega_e}\frac{\omega_f}{s+\omega_f}\right)}{1 + \left(K_1 + K_2\frac{1}{s}\right)\left(K_s \frac{\omega_e}{s+\omega_e}\frac{\omega_f}{s+\omega_f}\right)}$$

$$= \frac{K_1 K_s \omega_e \omega_f s + K_2 K_e \omega_e \omega_f}{s^3 + \left(\omega_e + \omega_f\right)s^2 + (1 + K_1 K_s)\omega_e \omega_f s + K_2 K_s \omega_e \omega_f} \tag{7.57}$$

In the design of PI controllers for the force-to-balance loop, the ITAE index can also be used for optimization. As shown in Table 7.2, the optimal transfer function of the force-to-balance control system is what is shown in Eq. (7.54).

According to the transfer function in drive mode, the PID parameters used for vibration amplitude control can be obtained. The various parameters in the hardware circuit of the PID controller can also be determined.

Similarly, the PI controller of force-to-balance circuit 2 can be designed by the above method as well. There is only a phase shift between the two demodulated signals and modulated signals used for force balance, while there is no difference in frequency or amplitude.

### 7.4.2   Angular Velocity Signal Detection Techniques

When there is an angular velocity input into the sensitive axis of the gyroscope, the Coriolis force will excite the sensitivity mode of the resonator, meaning the vibration of the resonator in sensitivity mode also contains angular velocity information.

According to the force-to-balance principle of the resonator, the signal amplitude in any position of the force-to-balance circuit of the resonator contains input angular velocity information. This means that angular velocity information can be extracted by demodulating the signal of the force-to-balance loop and the signal of the resonator in drive mode.

All PI demodulation and modulation branches of the force-to-balance loop have this kind of demodulation function. For a CVG, the drive mode frequency should be very close to the sensitivity mode frequency. Moreover, the resonator has a high mechanical quality factor, and the angular velocity signal of the gyroscope should be demodulated after a 90° drive signal phase shift. Therefore, the DC output from the PI-1 controller can be directly used as an angular velocity signal, as shown in Fig. 7.15.

**Fig. 7.15** Signal flow diagram of the angular velocity detection system

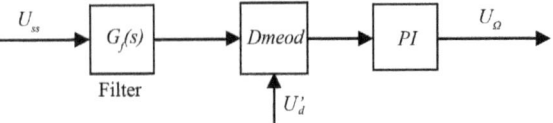

## 7.5 Introduction to Digital Circuits

### 7.5.1 Overview of Measurement and Control Circuits

According to the driving and detection characteristics of a CVG, its measurement and control circuit consists of three parts: a signal conditioning module, an analog digital signal interface module and a digital signal processing module. As shown in Fig. 7.16, a digital-to-analog hardware circuit is shown on the left side, comprising a signal conditioning module which in turn includes an amplifying circuit, a single-ended to differential circuit and an active low-pass filter. In the middle lies an interface between analog signals and digital signals that consists of a digital-to-analog converter and an analog-to-digital converter. On the right is a digital signal processing module composed of digital circuits [4].

A high-precision instrumentation amplifier is used to amplify voltage signals, while the low-pass filter is an active filter equipped with a precision operational amplifier. The above circuit has a relatively simple principle and is used in a wide variety of applications. A large amount of research has focused on this type of circuit, so it will not be discussed in detail here.

The hardware is designed to be combined with software for implementation of high-precision sinusoidal drive signals. The drive and detection loops are also synchronously demodulated using a digital method. In addition, other digital elements are added, including a digital PI controller, a digital modulator, a digital filter, a hardware circuit and logic controller and a number of error compensation units. Closed-loop drive control and angular velocity extraction are the core parts of the overall measurement and control system and have a key effect on the gyroscope's performance. Below is an introduction to the closed-loop drive control set

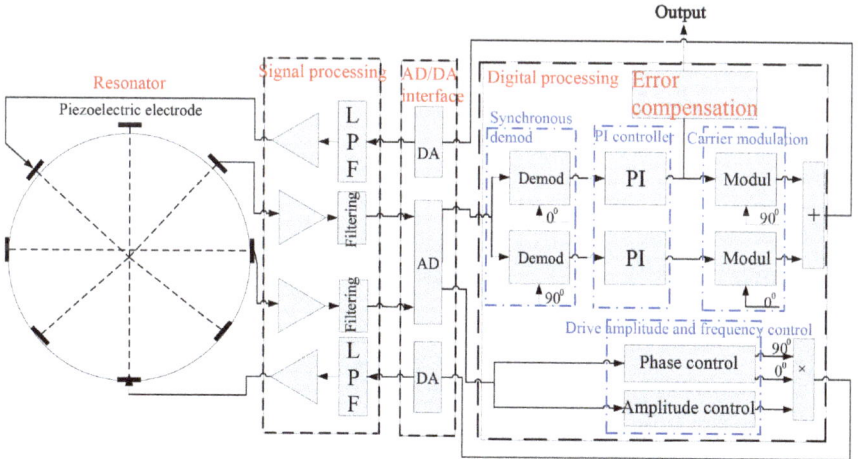

**Fig. 7.16** Block diagram of the measurement and control circuit of CVG

up and the closed-loop angular velocity detection scheme adopted in this paper. The former is used for closed-loop drive control, to ensure that the gyroscope works stably at the resonance point, while the latter is used for angular velocity extraction and calculation.

## 7.5.2   Principles of Closed-Loop Drive Control

Closed-loop drive control involves two aspects: vibration frequency control and vibration amplitude control. The closed-loop frequency control aims to ensure the frequency of the drive signals track any changes in the resonant frequency in the drive mode. This will cause the vibration amplitude to increase, improving gyroscope detection sensitivity; the purpose of amplitude control is to enable the resonator to vibrate at a constant amplitude, thus further improving stability.

From a kinetic perspective, the two modes of the CVG can be equivalent to a second-order spring-mass-damping system. The oscillator mass on the $x$- and $y$-axis are denoted by $m_x$ and $m_y$, respectively; $k_x$ and $D_x$ represent the $x$-axial elastic coefficient and damping coefficient of the gyroscope, respectively; $k_y$ and $D_y$ represent the $y$-axial elastic coefficient and damping coefficient, respectively. As a result, the 2D motion equation of the system can be written as:

$$\begin{bmatrix} m_x & 0 \\ 0 & m_y \end{bmatrix} \begin{bmatrix} \ddot{x} \\ \ddot{y} \end{bmatrix} + \begin{bmatrix} D_x & 0 \\ 0 & D_y \end{bmatrix} \begin{bmatrix} \dot{x} \\ \dot{y} \end{bmatrix} + \begin{bmatrix} k_x & 0 \\ 0 & k_y \end{bmatrix} \begin{bmatrix} x \\ y \end{bmatrix} = \begin{bmatrix} F_x \\ F_y \end{bmatrix} \tag{7.58}$$

where $F_x$ and $F_y$ represent the $x$-axial and $y$-axial driving forces; $F_x$ represents an external force applied on the drive shaft of the CVG; $F_y$ represents the Coriolis force coupled to the detection terminal.

When the drive shaft is excited and generates a driving force $F_x = F_d \sin \omega_d t$, the vibration displacement in the drive mode can be evaluated by:

$$x(t) = A_x \sin(\omega_d t - \varphi_d) \tag{7.59}$$

where

$$A_x = \frac{F_d/m}{\sqrt{(\omega_{dx}^2 - \omega_d^2)^2 + \omega_{dx}^2 \omega_d^2 / Q_x^2}} \tag{7.60}$$

$$\varphi_d = \arctan \frac{\omega_{dx} \omega_d}{(\omega_{dx}^2 - \omega_d^2) Q_x} \tag{7.61}$$

where $\omega_{dx} = \sqrt{k_x/m}$ represents the resonant frequency in drive mode; $Q_x = \sqrt{k_x m}/D_x$ represents the quality factor in drive mode.

The phase of the signal output from the drive end is $\varphi_d$. After demodulation, the phase of the vibration displacement signal can be obtained, as shown in Eq. (7.61). Due to the inherent delay within the lines and the phase delay of various filters in the circuit, the phase signal obtained after demodulation contains the phase $\varphi_d$ caused by the drive mode and an additional phase delay caused by all other factors. If the additional delay is $\varphi_{delay}$, the phase obtained after demodulation can be written as follows:

$$\varphi = \varphi_{delay} + \varphi_d \tag{7.62}$$

Equation (7.61) and (7.62) are solved simultaneously. The results are as follows:

$$\tan(\varphi_{delay} - \varphi) = \frac{\omega_{dx}\omega_d}{(\omega_{dx}^2 - \omega_d^2)Q_x} = \frac{(\omega_{dx}\omega_d)/\omega_{dx}^2}{Q_x(\omega_{dx}^2 - \omega_d^2)/\omega_{dx}^2} = \frac{1}{Q_x}\frac{\omega_d/\omega_{dx}}{1 - (\omega_d/\omega_{dx})^2} \tag{7.63}$$

It therefore follows that

$$\left(\frac{\omega_d}{\omega_{dx}}\right)^2 + \frac{1}{\tan(\varphi_{delay} - \varphi) \cdot Q_x} \cdot \left(\frac{\omega_d}{\omega_{dx}}\right) - 1 = 0 \tag{7.64}$$

According to the above-mentioned cause of $\varphi_{delay}$, its magnitude is relatively fixed, so it can be offset by hardware or software in an actual circuit. Thus, we can let $\varphi_{delay} = 0$, meaning

$$\frac{\omega_d}{\omega_{dx}} = \frac{-1/(\tan\varphi \cdot Q_x) + \sqrt{1/(\tan\varphi \cdot Q_x)^2 + 4}}{2} \tag{7.65}$$

Because $Q_x$ is generally equal to a value that is greater than one thousand, or even ten thousand, when $\varphi = \pi/2$, $\omega_d/\omega_{dx} = 1$. This means that no matter what $Q_x$ is equal to, when the driving frequency applied is equal to the resonant frequency in drive mode, the phase caused by the drive mode will be $\pi/2$. Therefore, it can be inversely inferred that when the phase caused by the drive mode is $\pi/2$, the driving frequency applied at this time is equal to the resonant frequency in drive mode, and the resonator will be in its resonant state. Therefore, phase control can be performed so that frequency $\omega_d$ of the voltage signal in drive mode will track the resonant frequency $\omega_{dx}$.

The above is the principle of closed-loop frequency control in drive mode. For amplitude control, the amplitude $A_x$ of the vibration displacement signal in drive mode is directly proportional to the driving force $F_d$. $F_d$ is generated by the piezo-electric plate in the form of piezoelectric transduction, so its magnitude is directly proportional to the sinusoidal voltage magnitude $U_{input}$ of the driving piezoelectric electrode. Therefore, closed-loop amplitude control can be performed by demodu-lating $F_d$ to judge the magnitude of the amplitude and by adjusting the driving voltage

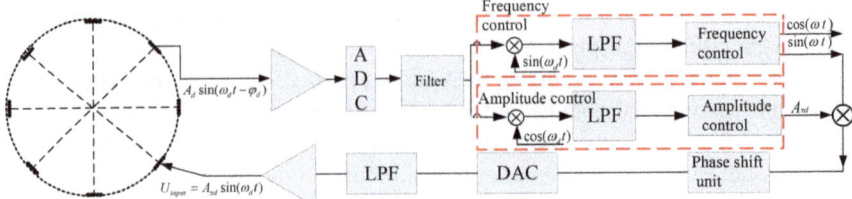

**Fig. 7.17** Principle of orthogonal drive control

$U_{\text{input}}$ to ensure the resonator vibrates at a constant amplitude. After the above analysis, we can reveal the implementation principle of closed-loop drive control, as shown in Fig. 7.17.

Quadrature demodulation can effectively extract a control target, making it easier to subsequently implement PI control, thus ensuring more effective control over the drive loop. The specific principle contains demodulation multiplication and low-pass filtering, as shown in Fig. 7.18. $\sin(\omega_d)$ and $\cos(\omega_d)$ are digital reference signals that change with the driving frequency but have a fixed initial phase.

This means that for the first loop,

$$A_d \sin(\omega_d t - \varphi_d) \sin(\omega_d t) = \frac{A_d \cos \varphi_d}{2} - \frac{A_d \cos(2\omega_d t - \varphi_d)}{2} \qquad (7.66)$$

After filtering, the following can be derived:

$$x = \frac{A_d \cos \varphi_d}{2} \qquad (7.67)$$

Then, for the second loop

$$A_d \sin(\omega_d t - \varphi_d) \cos(\omega_d t) = -\frac{A_d \sin \varphi_d}{2} + \frac{A_2 \sin(2\omega_d t - \varphi_d)}{2} \qquad (7.68)$$

After filtering, the following can be derived:

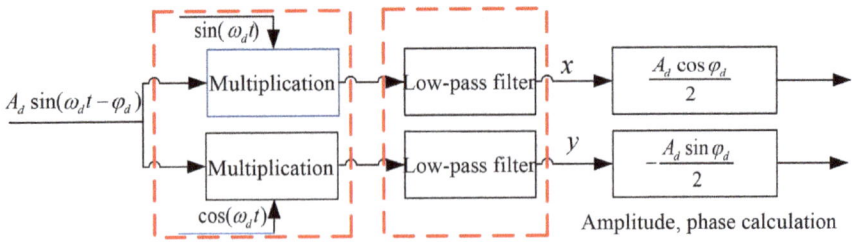

**Fig. 7.18** Schematic diagram of vibration amplitude and phase extraction

$$y = -\frac{A_d \sin \varphi_d}{2} \tag{7.69}$$

When the drive signal frequency is equal to the drive mode frequency, $\varphi_d = \pi/2$, so theoretically, $x = \frac{A_d \cos \varphi_d}{2} = 0$ while $y = -\frac{A_d}{2}$.

Therefore, for a PI controller used for frequency control, ideally, its control target value can be set to zero. When the closed loop enters a stable state, i.e. when $x = 0$, it can be considered that the resonator is in a resonant state. When this happens, amplitude control begins. If the current value is different from the target amplitude and an error is caused, the PI controller will generate a control variable to adjust the amplitude of the drive signal and enable the gyroscope to vibrate at constant amplitude.

When modal frequency drift is caused by external or internal factors, the drive signal frequency will no longer be equal to the mode frequency, thus leading to a change in $\varphi_d$, i.e., $x \neq 0$, with error $\Delta x$. The PI controller will generate a frequency control variable to adjust the drive signal frequency so that it can follow any change in modal frequency.

### 7.5.3 Principles of Closed-Loop Detection for Angular Velocity

For the vibration voltage signal that modulates the external angular velocity $\Omega_z$, the DC value obtained after filtering demodulation is directly proportional to the external angular velocity $\Omega_z$. The external angular velocity $\Omega_z$ can then be calculated based on the scale factor.

All detection signals are digitally demodulated by a demodulation setup that is similar to lock-in amplification. The structure is shown in Fig. 7.19.

The demodulation reference signals are $\sin \omega_{dx} t$ and $\cos \omega_{dx} t$, generated digitally, and the signal frequency $\omega_{dx}$ steadily follows the drive mode frequency. The fixed additional phase delay caused by the detection loop and the drive loop have been offset in advance, and the most basic detection loop principle is shown in Fig. 7.20.

The vibration signal $V_{sense1}$ at the output of the detector can be expressed as:

**Fig. 7.19** Demodulation of quadrature lock-in amplification

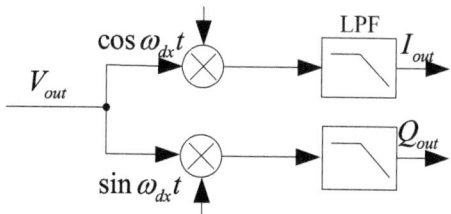

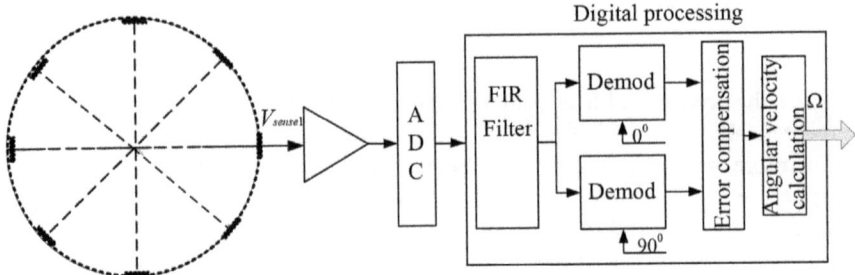

**Fig. 7.20** Functional block diagram for open-loop angular velocity detection

$$V_{sense1}(t) = K\Omega_z \sin(\omega_{dx}t + \varphi_{sense}) + V_Q \sin(\omega_{dx}t + \varphi_q) + V_I \sin(\omega_{dx}t + \varphi_{sense}) \tag{7.70}$$

where $V_Q \cos(\omega_{dx}t + \varphi_q)$ represents an orthogonal error signal; $\varphi_q$ represents a phase change caused by non-ideal factors or coupling between the sense axis and drive shaft; $V_I \sin(\omega_{dx}t + \varphi_{sense})$ represents a same-phase error signal; $K_z \sin(\omega_{dx}t + \varphi_{sense})$ represents an angular velocity coupling signal.

After $\cos\omega_{dx}t$ demodulation and filtering, the following can be derived:

$$I_{out} = K\Omega_z \sin\varphi_{sense} + V_Q \sin\varphi_q + V_I \sin\varphi_{sense} \tag{7.71}$$

After $\sin\omega_{dx}t$ demodulation and filtering, the following can be derived:

$$Q_{out} = -K\Omega_z \cos\varphi_{sense} - V_Q \cos\varphi_q - V_I \cos\varphi_{sense} \tag{7.72}$$

When the gyroscope is in its resonant state, it can be obtained from the preceding analysis that $\varphi_{sense} = \frac{\pi}{2}$, but ideally, $\varphi_q = \pi$ or $-\pi$, so the following can be derived respectively based on Eqs. (7.71) and (7.72): $I_{out} = K\Omega_z + V_I$, $Q_{out} = -V_Q$, where $I_{out}$ undergoes a linear change with angular velocity $\Omega_z$, while $Q_{out}$ is an orthogonal error signal component.

Owing to its poor anti-jamming properties and low precision, open-loop detection is not normally applied to gyroscope measurement and control systems. Close-loop detection is performed by means of force feedback, and the output quantity is compared with the input quantity to form a closed-loop control loop to improve the frequency response bandwidth, range and precision of the system. The basic principle of force feedback is shown in Fig. 7.21.

When there is an external angular velocity $\Omega_z$, the Coriolis force directly proportional to the external angular velocity deforms the sense axis of the resonator. The current vibration voltage signal can be sampled to generate a feedback voltage signal, which will then act on the moment device to produce an opposing force $F_b$, which will drive the sense axis of the resonator, and thus deform it. This deformation will be opposite to the deformation caused by the Coriolis force, and will suppress the vibration of the sense axis of the gyroscope and increase the dynamic response range

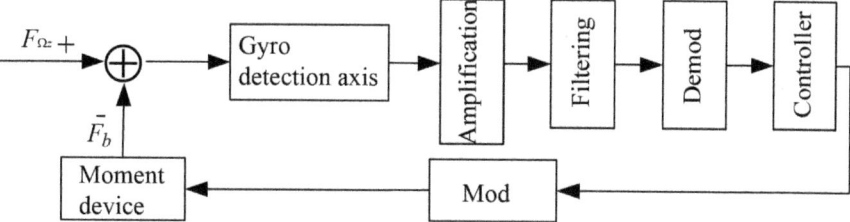

**Fig. 7.21** Force feedback-based closed-loop detection principle

of the sense axis. In this way, the bandwidth and detection precision of the gyroscope will be improved. Because the applied feedback force is equal to the Coriolis force, but lies in the opposite direction, the Coriolis force can be calculated by reading the magnitude of the feedback force. By conversion, the external angular velocity can then be evaluated.

### 7.5.4 Algorithm Flow Diagrams for Drive and Detection Loops

The digital control circuit software of the CVG is used for demodulation, filtering, PI control, drive signal generation and angular velocity calculation. When a calculation is carried out in the timer interrupt service routine, there is a slight difference in the software flow between the drive loop and the detection loop. The software flow diagram for the drive loop is shown in Fig. 7.22a while that for the detection loop is shown in Fig. 7.22b.

Firstly, after the software is initiated, various functions and hardware modules are initialized. Then, the timer is used to perform a sweep check. As the drive frequency approaches, the closed-loop drive control will be initiated. The detection loop will not work at this time. The control procedure for the detection loop will wait until the closed-loop control enters a steady state; this is the most time-consuming part of the gyroscope start-up process. After the drive loop starts, the closed-loop control of the detection loop will begin. After a series of mathematical operations, such as reading the AD data in the detection signal output channel, demodulation, PI control and orthogonal modulation, a force feedback signal will be generated to balance the vibration of the sense axis. When this vibration is suppressed to a certain level, it means that the gyroscope can be fully started. The sum of these two time durations is equal to the start time of the gyroscope. During timer interruption, the timer control will be reloaded to repeat the next control flow.

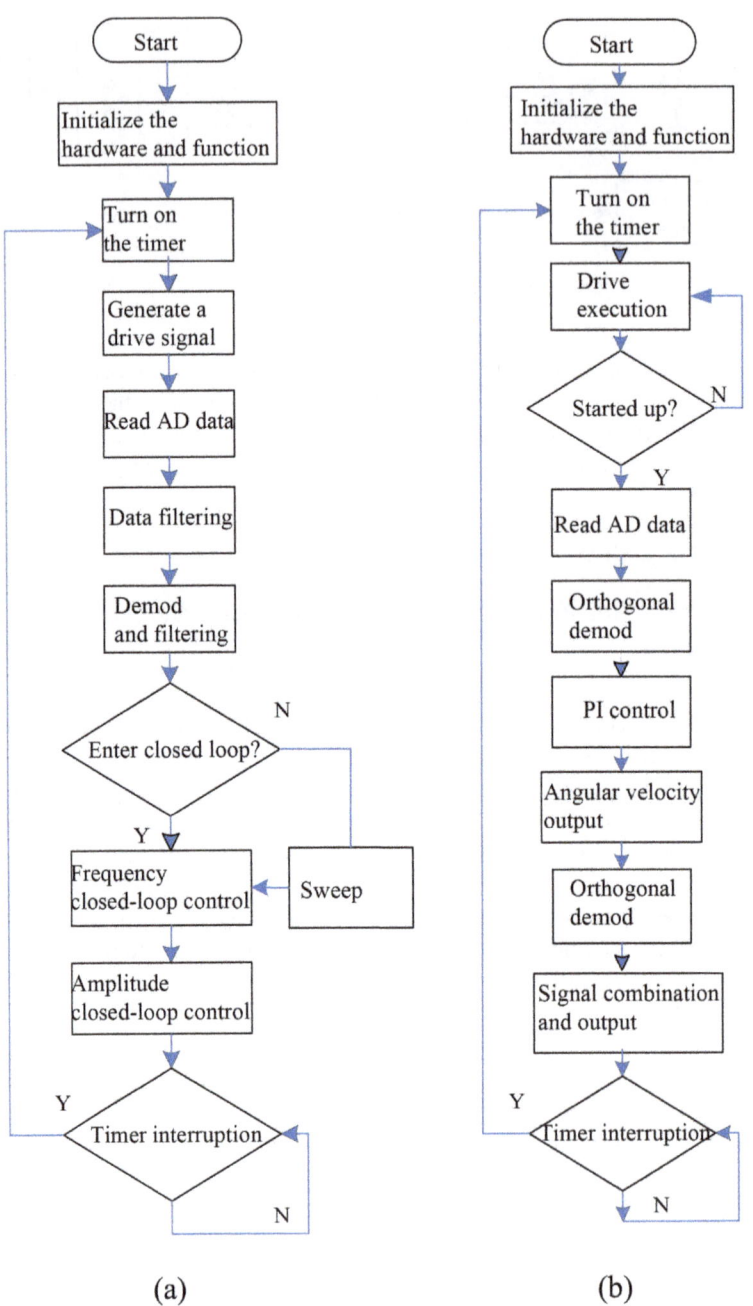

**Fig. 7.22** Flow diagrams for the drive loop and detection loop. **a** Algorithm flow diagram for drive loop; **b** algorithm flow diagram for detection loop

# References

1. Yi, T. (2011). *Research on the key technology of cup-shaped wave gyros.* National University of Defense Technology.
2. Wang, X., Wu, W., Luo, B., Fang, Z., Li, Y., & Jiang, Q. (2011). Force to rebalance control of HRG and suppression of its errors on the basis of FPGA. *Sensors, 11*(12), 11761–11773.
3. Cui, J., Guo, Z. Y., Zhao, Q. C., Yang, Z. C., Hao, Y. L., & Yan, G. Z. (2011). Force rebalance controller synthesis for a micromachined vibratory gyroscope based on sensitivity margin specifications. *Journal of Microelectromechanical Systems, 20*(6), 1382–1394.
4. Di, X. (2011). *Research on the digital measurement and control technology of cup-shaped wave gyroscopes.* National University of Defense Technology.

# Chapter 8
# Error Mechanism and Compensation in CVGs

Ideally, when the input angular velocity is zero, the gyroscope output will also be zero. However, considering unavoidable manufacturing errors, material defects, structural stresses and the non-ideal nature of drive circuit components, the gyroscope output is never actually equal to zero, even with no angular velocity input. The output at this moment is known as the bias of the gyroscope. Constant bias can be eliminated by software compensation, to avoid any impact on the navigation system. However, the intrinsic instability of bias results in an incorrect angular velocity being output, which seriously affects the accuracy of the navigation and attitude control system. Because of the significant impact of ambient temperature on the material characteristics and structural stress of the CVG, as well as the effect of random noise from the mechanical structure and the measurement and control circuit, the bias will vary with time and ambient temperature [1]. Under static conditions, the long-term steady-state output of the CVG is a random, stable process. Bias stability refers to the degree of dispersion of the CVG output around its mean value (bias), reflecting the level of changes and fluctuations in gyroscope output when in a static state. Generally speaking, this is given in °/h. Figure 8.1 shows a typical bias output curve of the CVG, which is mainly composed of bias drift and random noise.

There are many factors that cause gyroscope output drift. For a high-precision inertial navigation gyroscope, the main causes of drift include the inherent imperfections in gyroscope principles, structure and technology. Other causes are interference from linear and angular motion, but these external factors would not exist without internal factors.

© National Defense Industry Press 2021

X. Wu et al., *Cylindrical Vibratory Gyroscope*, Springer Tracts in Mechanical Engineering, https://doi.org/10.1007/978-981-16-2726-2_8

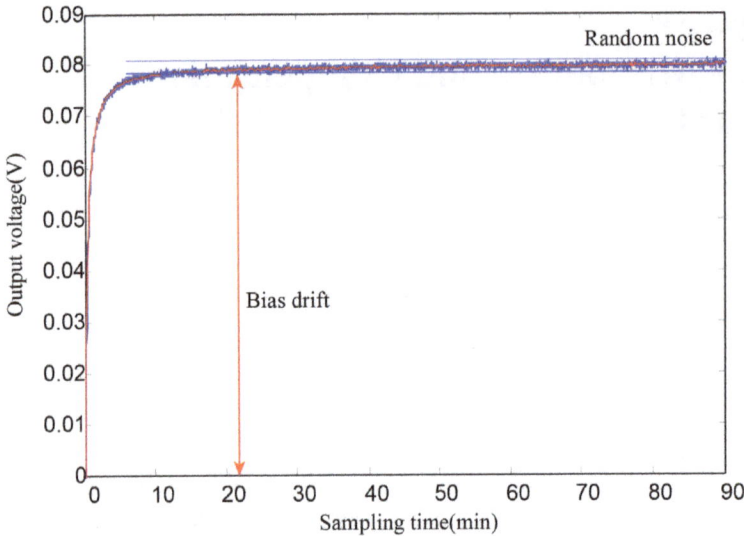

**Fig. 8.1**  Typical bias output curve

## 8.1  Main Error Sources for CVGs

### 8.1.1  Defects in Resonator Parameters

#### I.  Material Defects

An ideal resonator material should contain evenly distributed crystalline grains and be homogeneous all directions in terms of density, stiffness and damping coefficient. But owing to congenital material defects, and because of new defects caused in the machining process, the resonator material parameters actually become a function of the circumferential angle θ [2].

Let homogeneous resonator density be $\rho_0$. The following is its variable density function when there are errors present:

$$\rho = \rho_0 + \delta\rho(\theta) \tag{8.1}$$

where $\delta\rho(\theta)$ represents material errors caused by non-ideal factors, and the distribution is related to circumferential angle θ.

Because a continuous non-ideal density function is not conducive to direct analysis or calculation, it can be discretized, by means of Fourier transform, as follows:

$$\delta\rho(\theta) = \sum_{k=-\infty}^{+\infty} \{\delta\rho\}_k e^{ik\theta} \tag{8.2}$$

where $k$ represents a natural number, and $\{\delta\rho_k\}$ represents the $k$th-order density component of the Fourier expansion coefficient.

Similarly, non-ideal stiffness can be discretized as follows

$$\delta K(\theta) = \sum_{k=-\infty}^{+\infty} \{\delta K\}_k e^{ik\theta} \tag{8.3}$$

where $\{\delta K_k\}$ represents the $k$th-order stiffness component of the Fourier expansion coefficient. This analysis can be facilitated by discretizing continuous material errors into discrete errors [3].

In addition to density and stiffness defects, the resonator has another important type of defect: residual stress. A residual stress refers to a state of stress balance maintained within a piece after machining is finished. The residual stress has a very significant impact on high-precision components and changes slowly over time. This slow change causes a rebalancing of the internal stress of the piece, thus deforming it. Residual stress is the most critical factor affecting the form and positional accuracy of the resonator, so it must be suppressed by a strict process of thermal treatment. The following are the main causes of residual internal stress:

(1)   The inherent internal residual internal stress of the material. Crystalline grains form in different sizes and directions during the melting and solidification process of a metal material. These crystalline grains compress one another, leading to internal stress within the material.
(2)   Residual internal stress caused in the machining process. During the process of resonator machining, the cutter interacts with the rough resonator, disconnecting some of the material from the bar stock. After machining, the structural surface layer is compressed by the cutter, bringing a large residual internal stress into being.
(3)   Residual internal stress caused by changes in temperature. Because the metal material used for resonator production is not a perfect elastomer, elastic transition-caused plastic deformation often causes a residual internal stress when the temperature rises or falls.

## II.   Geometric Errors

The accuracy of the CVG resonator needs to be of the order of micrometers, while its shell should be below 0.5 mm at the thinnest point; in other words, the shell is a highly deformable, thin-walled part, meaning significant geometric errors are inevitable. Machining precision detection is the primary condition for analyzing and suppressing geometric errors. For the resonator, its roundness and uniformity of wall thickness matter the most.

Because the resonator vibrates in the form of a circumferential standing wave, and since the vibration in this mode is affected by periodic mass distribution, the unordered geometric shape of the resonator can be decomposed in order to observe its internal components. If the wall thickness of the resonator is decomposed into multi-order discrete functions, the wall thickness errors can be expressed as a harmonic function.

$$\delta h(\theta) = h_0 + \sum_{k=1}^{\infty} (a_k \cos n\theta + b_k \sin n\theta) \tag{8.4}$$

where $h_0$ represents the wall thickness of the resonator under ideal conditions, and $a_k$ and $b_k$ represent the sinusoidal and cosinusoidal harmonic error components, respectively.

Taking a resonator with a wall thickness error of about 1 μm as an example, the first 20 harmonics of the resonator are fitted by means of the least squares method. The results are shown in Fig. 8.2. It should be noted that the higher the harmonic order, the higher the fitting precision, but that the calculation complexity will also increase. It can be seen that the maximum component of the resonator is the second harmonic component (coefficient: 0.135), indicating that the main machining deformation of the resonator is elliptical (Fig. 8.3a). Furthermore, the fourth harmonic component coefficient of the resonator is 0.047, primarily affecting the frequency split of the resonator due to the largest vibration mass difference between its drive and sense modes (Fig. 8.3b). Other harmonic components also exist, but owing to their small coefficient and lack of direct contributions to the main performance indexes of the resonator, they are not all listed here.

Frequency mismatch is primarily caused by the fourth harmonic density/geometric errors. Given uniform density, the frequency split $\Delta\omega$ caused by fourth harmonic errors can be estimated by

$$\Delta\omega \cong \omega \frac{m_4}{m_0} \cong \omega \frac{h_4}{h_0} \tag{8.5}$$

where $m_4$ represents the equivalent vibration mass of the fourth harmonic; $m_0$ represents the total vibration mass of the resonator; $h_4$ represents the non-uniformity of the

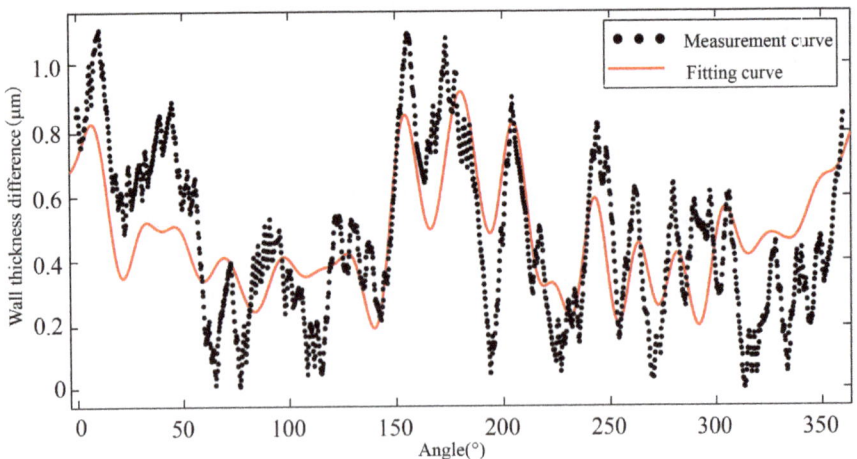

**Fig. 8.2** Harmonic fitting of thickness errors

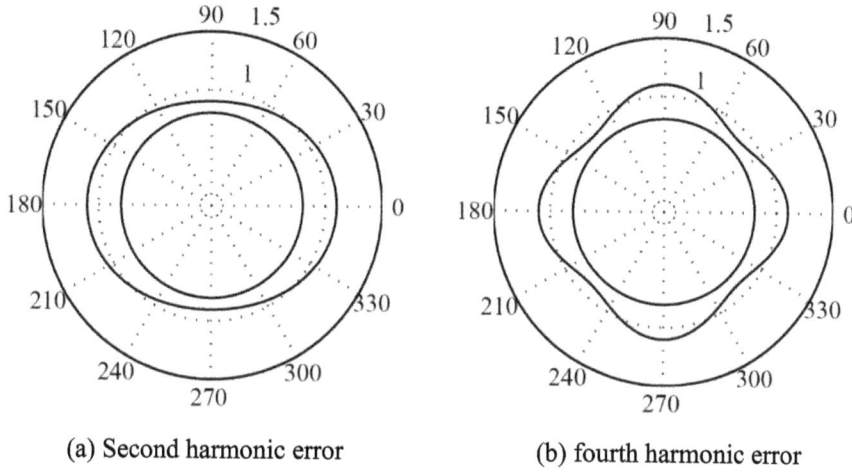

(a) Second harmonic error                          (b) fourth harmonic error

**Fig. 8.3**  Geometric error represented by harmonic components

wall thickness caused by the fourth harmonic; and $h_0$ represents the wall thickness of the resonator.

It therefore follows that, if the frequency split of the resonator has to be controlled below a certain ratio, the wall thickness error has to also be controlled below that same ratio. Considering that the working frequency of a resonator with a diameter of 20–30 mm is generally around 4000 Hz, the wall thickness of the resonant ring is set to 1–2 mm. If the initial machining error must be controlled within 1 Hz, the fourth harmonic error of the wall thickness has to be below 0.5 μm.

## 8.1.2  Temperature Errors

### I.   Impact of Temperature on the Elastic Modulus of the Resonator

A CVG resonator is generally made of metal. Normally, temperature has a significant impact on the elastic modulus of metal materials. Most solids expand with increasing temperature, which is also accompanied by a rise in material volume and a decrease in atomic binding forces. Intuitively, the material will become increasingly soft, meaning there is a close relationship between the elastic model and temperature. Usually, the elastic modulus temperature coefficient $\beta_E(1/°C)$ represents the variation of the elastic modulus with a change in temperature of 1 °C.

$$\beta_E = \frac{dE}{EdT} \tag{8.6}$$

II.　**Impact of Temperature on the Poisson's Ratio of the Resonator**

The Poisson's ratio refers to the ratio of the transverse strain of a material to its longitudinal strain. It is also known as the lateral deformation coefficient of the material. Taking a rectangular pole as an example, when there is a stress $\sigma_x$ along the length of the pole, it will extend longitudinally. Let the extension be $\varepsilon_x$. The pole will contract transversely, and the contraction in the $y$ and $z$ directions is given as

$$\varepsilon_y = \frac{l_y' - l_y}{l_y} = -\frac{\Delta l_y}{l_y} \tag{8.7}$$

$$\varepsilon_z = \frac{l_z' - l_z}{l_z} = -\frac{\Delta l_z}{l_z} \tag{8.8}$$

Therefore, the Poisson's ratio of materials is defined as:

$$\mu = \left| \frac{\varepsilon_y}{\varepsilon_x} \right| = \left| \frac{\varepsilon_z}{\varepsilon_x} \right| \tag{8.9}$$

When a CVG is opearting, the inner heat and ambient temperature changes cause the temperature of the resonator to vary in different locations, forming a temperature gradient. Therefore, different thermal stresses are generated in different elements of the resonator. Thus, the values of $\varepsilon_y$ and $\varepsilon_z$ are not fixed, but change slightly with temperature, meaning the Poisson's ratio $\mu$ is not a fixed value either.

III.　**Impact of Temperature on the Material Density of the Resonator**

Let the volume of the resonator per unit mass be $V$, so the material density of the resonator becomes:

$$\rho = \frac{m}{V} \tag{8.10}$$

When the temperature of the resonator changes, its volume per unit mass changes to:

$$V(T) = V_0(1 + \beta \Delta T) \tag{8.11}$$

where $\beta$ represents the thermal expansion coefficient of the material.

　　Equation (8.11) is substituted into Eq. (8.10), revealing the relationship between the material density of the resonator and its temperature as follows:

$$\rho = \frac{m}{V_0(1 + \beta \Delta T)} \tag{8.12}$$

　　A CVG resonator is generally made of a metal material whose thermal expansion coefficient $\beta > 0$. According to Eq. (8.12), the material density of the resonator decreases with increasing temperature. When the resonator has a large temperature gradient, the linear expansion of its geometric dimensions will be uneven, deflecting the center of mass of the resonator and causing output drift on the CVG.

## 8.2 Error Model of CVGs

Gyroscope error models are divided into two types: determinate error models and random error models. Of these, the former can be further divided into disturbance models (for parameter variations in the sensitive physical model) and environmental models (for sensitive environmental interference), while the latter refers to uncertain factors and random drift.

### 8.2.1 Determinate Error Model

#### I. Physical Parameter Disturbance Model

For a CVG, the parameter variation in the physical model is generally caused by the non-ideality of the resonator structure; manufacturing defects may deflect the rigid axis from the drive and detection electrodes, causing the driving force to produce components in the non-driving direction, facilitating the coupling among different modes and leading to output errors. In the drive circuit of the gyroscope, the manufacturing errors caused by the electronic devices related to the control loop affect the performance of the drive circuit, thereby changing the parameters of the system. In other words, imperfect structural fabrication causes the largest parameter disturbance errors, including mode frequency, drive frequency, stiffness and calibration factors. The change in physical model parameters causes gyroscope bias $B$, axis misalignment error $O_A$, cross-coupling error $O_C$ and hysteresis error $O_H$.

Bias $B$ is generally divided into preheating bias $B_T^{(t)}$ and operating bias $B_O$. Their relationship can be expressed as:

$$B = B_T^{(t)} + B_O \tag{8.13}$$

where preheating bias $B_T^{(t)}$ can be described with the following equation:

$$B_T^{(t)} = B_T(1 - e^{\frac{-t}{T_B}}) \tag{8.14}$$

where $B_T$ represents the preheating bias coefficient and $T_B$ represents the preheating time constant.

Axis misalignment error $O_A$ is a type of error caused by the misalignment between the actual input axis and the reference input axis under zero input conditions. In general, this is described by the phase difference angle $\alpha_M$ or the direction cosine matrix $A_M$.

Cross-coupling error $O_C$ refers to an output error caused by the input perpendicular to the input reference axis and sensitive to the gyroscope. This can be expressed as

$$O_C(\Omega_C) = O_H \cdot \Omega_C \qquad (8.15)$$

where $O_C$ represents the cross-coupling error coefficient, and $\Omega_C$ represents the angular velocity acting on the non-input reference axis and the sensitive axis.

Hysteresis error $O_H$, related to the bandwidth of the system, refers to the maximum error of the upper and lower bounds corresponding to the instantaneous hysteresis of the measured value. In general, it is equivalent to the input angular velocity bias.

II.  **Environmental Error Model**

The environmental factors affecting gyroscopes primarily include external acceleration, external angular acceleration, ambient temperature, and pressure; for a packaged CVG, the impact of pressure can be ignored, so the first three factors can be considered to be the primary causes of the environment-sensitive drift $E$ of the CVG.

Environment-sensitive drift can be described as

$$E = E_{\dot{\Omega}} + E_G + E_{GG} + E_T \Delta T + E_{\nabla T} \nabla T \qquad (8.16)$$

where $E$ represents angular acceleration-sensitive drift; $E_G$ represents acceleration-sensitive drift; $E_{GG}$ represents acceleration square-sensitive drift; $E_T \Delta T$ represents temperature-sensitive drift; $E_{\nabla T} \Delta T$ represents temperature gradient-sensitive drift.

The model for temperature-sensitive drift $E_T \Delta T$ is generally determined by experimental fitting. Its second-order model is given as follows

$$E_T \Delta T = E_{T1}(T - T_{ref}) + E_{T2}(T - T_{ref})^2 \qquad (8.17)$$

where $T$ represents the actual internal temperature of the gyroscope; $T_{ref}$ represents the reference temperature; $E_{T1}$ and $E_{T2}$ represent the temperature sensitive drift coefficient.

### 8.2.2  Random Error Model

Random gyroscopic drift is usually fitted using an ARMA model; it is an important indicator for measuring gyroscopic precision and one of the major error sources for an inertial navigation system. To achieve a reduction in random gyroscopic drift, a common method is to create a random error model and perform Kalman filtering.

To sum up, gyroscope error sources can be described by Fig. 8.4.

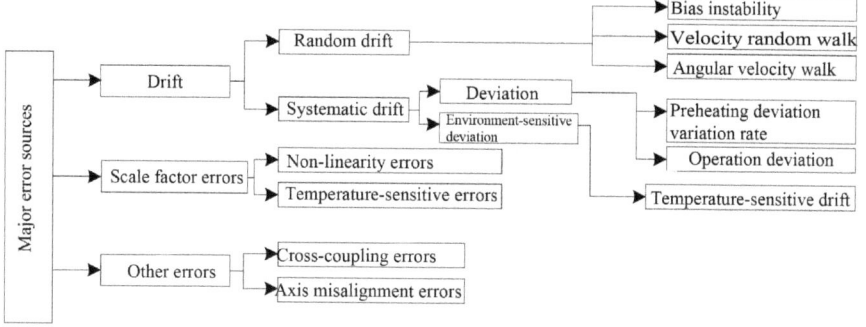

**Fig. 8.4**  Gyroscope error sources

## 8.3  Methods for Improving the Temperature Stability of CVGs

In terms of the characteristics of the CVG, there are three methods to choose from for improving temperature stability; these can be described as follows:

(1)  Improving the machining precision and selecting a material with a high quality factor. Machining errors of the resonator and the non-ideality of the material are the main causes of temperature drift in the CVG. Therefore, in order to improve the temperature performance of the gyroscope, we must raise the machining precision of the resonator, the uniformity of the resonator material and the thermal stability of the material to a very high level.

(2)  Gyroscope temperature control. Some hardware measures can be taken to keep the internal temperature of the gyroscope at a constant level. The usual method is to add a temperature control circuit inside the gyroscope. The temperature control circuit is generally composed of a heating or cooling device, a temperature sensor and a control circuit. The temperature control circuit is used to continuously correct the internal temperature of the gyroscope, enabling it to work at a constant temperature, thus greatly improving stability. However, the temperature control system increases the volume and weight of the gyroscope, while also disturbing the ambient temperature of the CVG. Before the temperature control circuit reaches its stable state, the gyroscope will drift unstably for a while, extending its stabilization time.

(3)  Software compensation for temperature errors. The premise of using software to compensate for temperature errors is based on the fact that the characteristics of the gyroscope change regularly with temperature, and that this process shows good repeatability [4]. Using this method, we can perform a temperature experiment to measure the bias at different temperatures and build a gyroscope bias temperature model. This mathematical model can then be imbedded into the control chip to compensate for gyroscope output in real time. The hardware

is easy to implement, but a temperature sensor is generally required, increasing the complexity of the system and the cost of the gyroscope.

## 8.4  Temperature Error Compensation for CVGs

### 8.4.1  Method of Temperature Error Compensation

The present section introduces a software compensation method that can be used to improve the temperature performance of the CVG [5]. The traditional temperature compensation method involves creating a mathematical model for gyroscope output and temperature by testing the gyroscope temperature. The model is then written into the temperature compensation circuit, and the current working temperature of the gyroscope measured by means of the temperature sensor to determine the factor required for temperature compensation. This can effectively reduce the tempera-ture drift of the gyroscope, but the problem of temperature hysteresis still remains, meaning the accuracy of temperature compensation is not very high.

The temperature hysteresis of the gyroscope is shown in Fig. 8.5. It is primarily caused by a temperature measurement error. Because there is a certain hysteresis between the temperature measured using the sensor and the true temperature of the gyroscope resonator, the curves showing the rise and fall of temperature between the gyroscope output and the measured temperature do not overlap. To overcome the hysteresis of temperature measurement, the frequency measurement method is

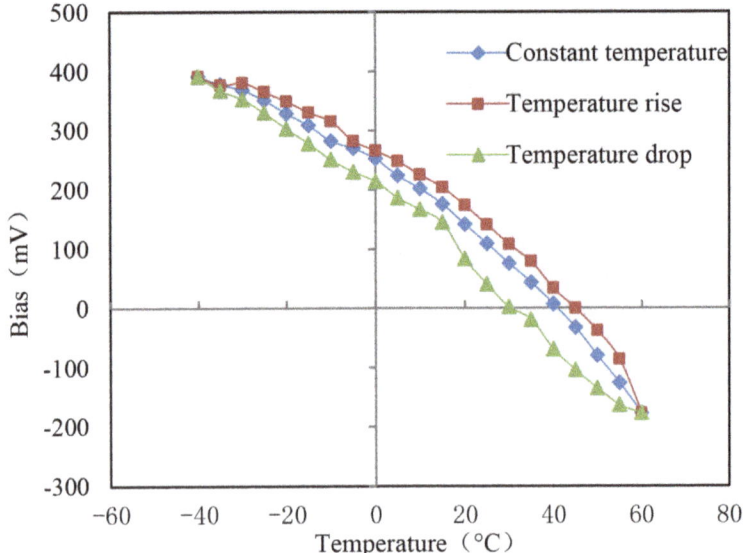

**Fig. 8.5** Temperature hysteresis within the gyroscope

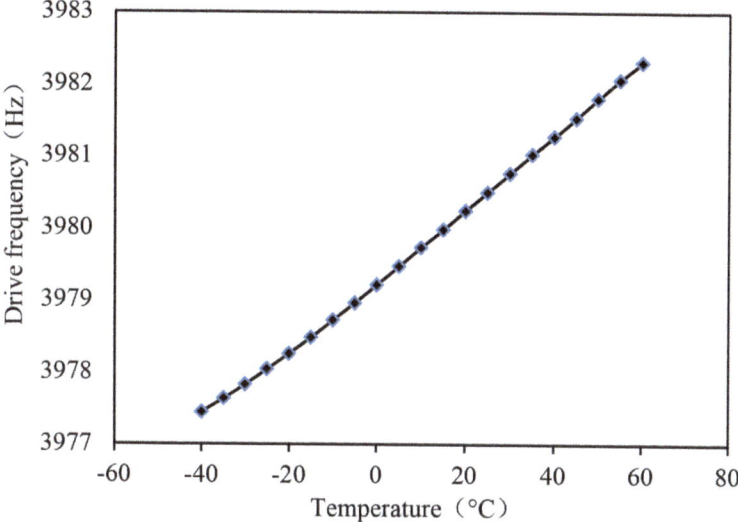

**Fig. 8.6** Resonator frequency-temperature curve

used for gyroscope temperature calibration. The resonant frequency is an inherent characteristic of the resonator and able to reflect its actual temperature in real time. The relationship between the resonant frequency and the temperature of the CVG is shown in Fig. 8.6. This shows that there is effectively a linear relationship between the resonant frequency and the temperature of the CVG.

Figure 8.7 shows the changing curves of the gyroscope output and drive frequency during consecutive temperature variations. The figure shows that both drive frequency and bias change with temperature, and that the inflection point of the drive frequency curve is very close to that of the bias curve on the time axis, indicating a very small delay between the drive frequency and bias, with both changing almost simultaneously. This illustrates that the drive frequency can reveal the bias drift caused by temperature changes in time. As a result, the drive frequency can be selected as a compensation parameter of the CVG for real-time gyroscope output compensation.

## 8.4.2 Composition of Temperature Compensation System

Use of a bias-frequency model for CVG compensation can avoid the necessity of a temperature sensor and reduces the complexity of the system, thus improving the real-time compensation performance. However, this method also has certain limitations. For example, it is difficult to implement high-precision frequency measurements. Furthermore, its resonant frequency is slightly changed by other factors as the gyroscope operates. It is easy to use a temperature sensor for temperature compensation,

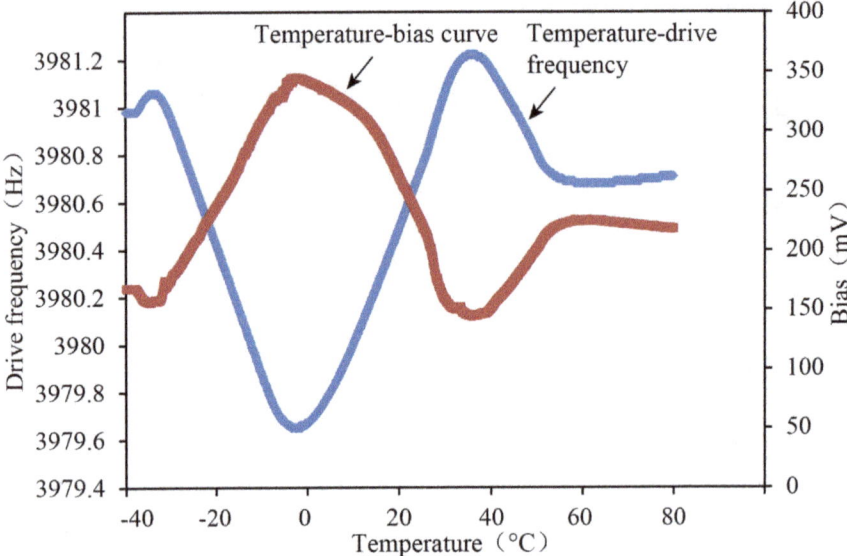

**Fig. 8.7** Gyroscope bias-frequency curve

and in engineering terms, this is relatively easy to implement. Moreover, the temperature of the resonator is not affected by its operating state, which is also of practical value.

The following is the general idea of temperature compensation: a temperature model is built for the CVG as part of a temperature experiment. By means of polynomial fitting, a mathematical expression can be established for the bias, temperature or drive frequency of the CVG throughout the whole temperature spectrum; this mathematical expression is written into the MCU memory. The temperature or drive frequency of the gyroscope is measured in real time during the compensation process. The compensation factor is calculated based on the mathematical expression in the MCU, and a gyroscope output signal can be obtained after compensation by subtracting the compensation factor from the gyroscope output.

The structure of the temperature compensation system is shown in Fig. 8.8. The temperature data measured with the temperature sensor or the frequency signal collected from the gyroscope drive circuit is imported into the MCU. Thanks to its internal A/D and D/A modules, the MCU then outputs the calculated compensation factor after conversion through the D/A. Post-compensation output is the bias signal minus the compensation factor output from the MCU. The MCU is the core of the whole system and processes all required data.

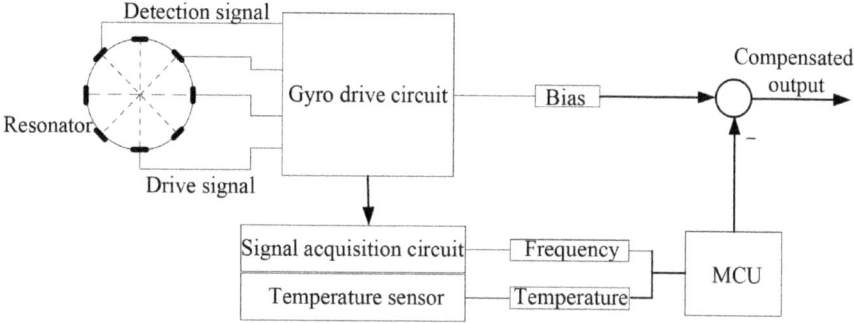

**Fig. 8.8**  Overall structure of the temperature compensation system

### 8.4.3 Hardware Implementation for Temperature Compensation System

For the temperature compensation circuit, its hardware includes the MCU peripheral circuit design, temperature sensor circuit design and compensation output circuit design.

(1)  MCU peripheral circuit design

The MCU is the core of the entire system, and its main function is reading, processing and outputting temperature sensor data. The design of the C8051 MCU is taken as an example. Its pin configuration and peripheral circuit are shown in Fig. 8.9.

In Fig. 8.9, MCU P0.2, P0.3 and P0.6 are configured as SPI interfaces, corresponding to SPI SCK, MISO and MOSI; these are used for the communication between the MCU and the temperature sensor. P0.4 and P0.5 are configured as serial communication interfaces, and used for serial communication with the upstream computer when necessary. C2CK and C2D are on-chip Silicon Labs second-line (C2) programming interfaces, used for non-intrusive (i.e. not occupying on-chip resources), full-speed and in-system debugging of the MCU. P0.0 and P0.1 are configured as DA outlets of the MCU to output the calculated compensation factor. The two DA outlets output a positive and a negative compensation factor, respectively. The internal oscillator of the MCU is used in order to reduce circuit power dissipation.

(2)  Temperature sensor circuit design

The temperature sensor is simple to use. Apart from the power supply and ground wire, the only steps required are to connect the data transmission pin DOUT and clock pin SCLK with the SPI SCK and MISO. The DIN pin and CS pin of the temperature sensor are earthed, meaning the temperature sensor is always in an active state, as shown in Fig. 8.10.

According to the foregoing analysis, there is some difference between the temperature measured using the temperature sensor and the actual temperature of the resonator. The temperature difference causes the hysteresis of bias temperature.

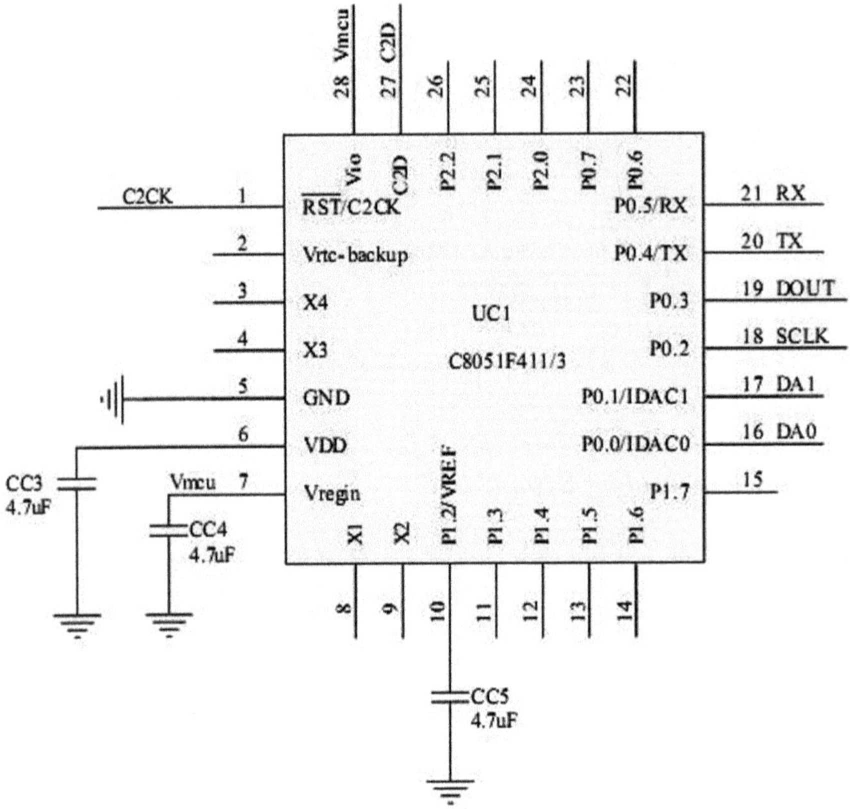

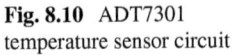

Fig. 8.9  C8051F411 MCU pin configuration

Fig. 8.10  ADT7301
temperature sensor circuit

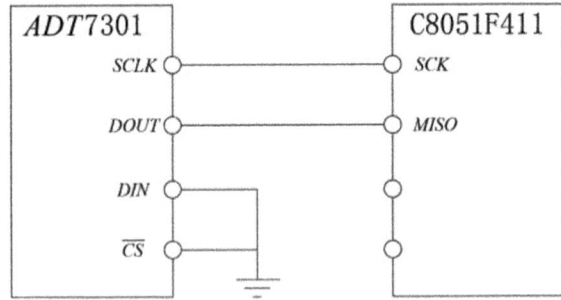

Temperature measurement accuracy is a key factor for the accuracy of temperature compensation. To reduce hysteresis and improve the compensation effect, the
temperature sensor should be located in a suitable location, so that the temperature
measured with the temperature sensor is as close as possible to the actual temperature
of the resonator, as shown in Fig. 8.11.

**Fig. 8.11** Temperature sensor installation diagram

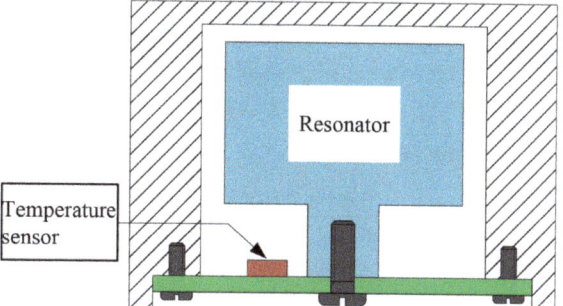

(3)  Compensation output circuit design

The main function of the compensation output circuit is to subtract the compensation factor output by the MCU from the gyroscope output and then filter it as the final output. The schematic circuit diagram is shown in Fig. 8.12.

In Fig. 8.12, DA0 and DA1 are DA outlets of the MCU. Because both DA outlets in C8051F411 MCU are current-type, the analog output represents current, while the output of the gyroscope represents voltage. Therefore, the DA output of the MCU needs to be transformed into voltage first. In this design, two precision 1 K resistors comprising the two loops of the DA current output are transformed into voltage output. The inverting input terminals of operational amplifiers 1 and 2 are short-circuited with the output terminal. They do not play an amplification role, but form a voltage follower circuit based on the properties of the operational amplifier. This includes "a virtual short circuit and a virtual open circuit", making the output voltage of the amplifier equal to the voltage of the input terminal, i.e., the offset voltage. Moreover, the resistance between them is infinite, meaning the compensation voltage will not be divided. The negative compensation factor and gyroscope output

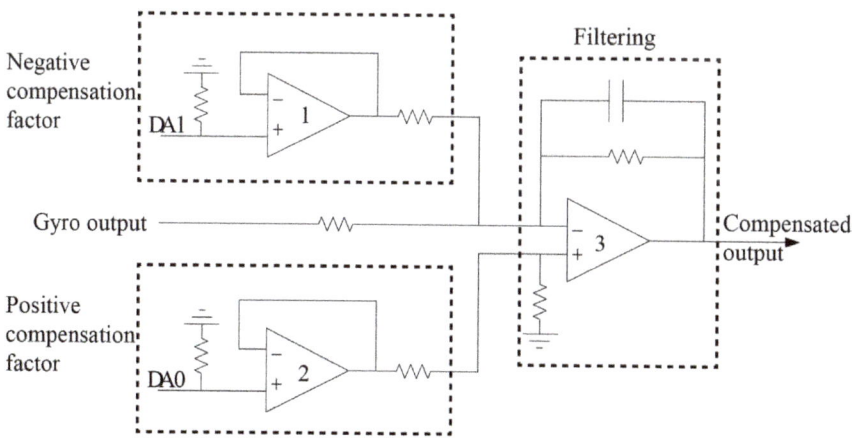

**Fig. 8.12** Compensation output circuit

are connected to the inverting input terminal of operational amplifier 3, while the positive compensation factor is connected to the same input terminal. Operational amplifier 3 also superimposes the three input loops over one another in order to filter them, obtaining a post-compensation output.

(4)  Routine design

The following is the basic idea of the routine: first, the temperature value is read from the sensor and the compensation factor is calculated based on the temperature drift model. This value is imported into a DA translation register in the MCU. If it is positive, it is imported into the DA0 translation register. If it is negative, it is imported into the DA1 translation register. It is then output after DA conversion.

Within the routine, a timer is used to control the output interval of the compensation factor. Owing to the slight difference in the temperature value read each time, the calculated compensation factor can also vary from time to time. If the output interval is too short, this amounts to adding noise to the gyroscope output, meaning the output interval should be slightly longer. The accuracy of the temperature sensor is not very high, which means that, in order to measure the temperature more accurately, the average of 100 continuously measured temperature values is taken as the current temperature value.

The routine flow diagram is shown in Fig. 8.13.

## 8.4.4  Temperature Compensation System Based on the Bias Frequency Model

(1)  Frequency measurement method

The main issue of the bias-frequency model-based temperature compensation system lies in how to accurately measure the frequency. Current frequency measurement methods include the direct frequency measurement method, the direct circumference measurement method and the equal-precision frequency measurement method. Direct frequency measurement refers to $N$ repeated changes in the signal to be measured within $T$, a specified time interval. The signal frequency $f_x = N/T$. The frequency measurement method has a high accuracy for high frequency measurements, but a low accuracy for low frequency measurements. Direct circumference measurement refers to the $N$ standard high-frequency signals $f_s$ that are detected within one cycle of the measured signals. In this case, the frequency $f_x = f_s/N$. The circumference measurement method has high accuracy for lower frequency measurements, but low accuracy for high frequency measurements. The equal-precision frequency measurement method has constant, high measurement accuracy within the entire frequency spectrum.

The principle of equal-precision frequency measurement is shown in Fig. 8.14.

In Fig. 8.14, the preset gating signal is a pulse with a width of $T_{pr}$. CNT1 and CNT2 are two controllable counters. A standard frequency signal has a frequency of

**Fig. 8.13** Bias-temperature
compensation routine flow
diagram

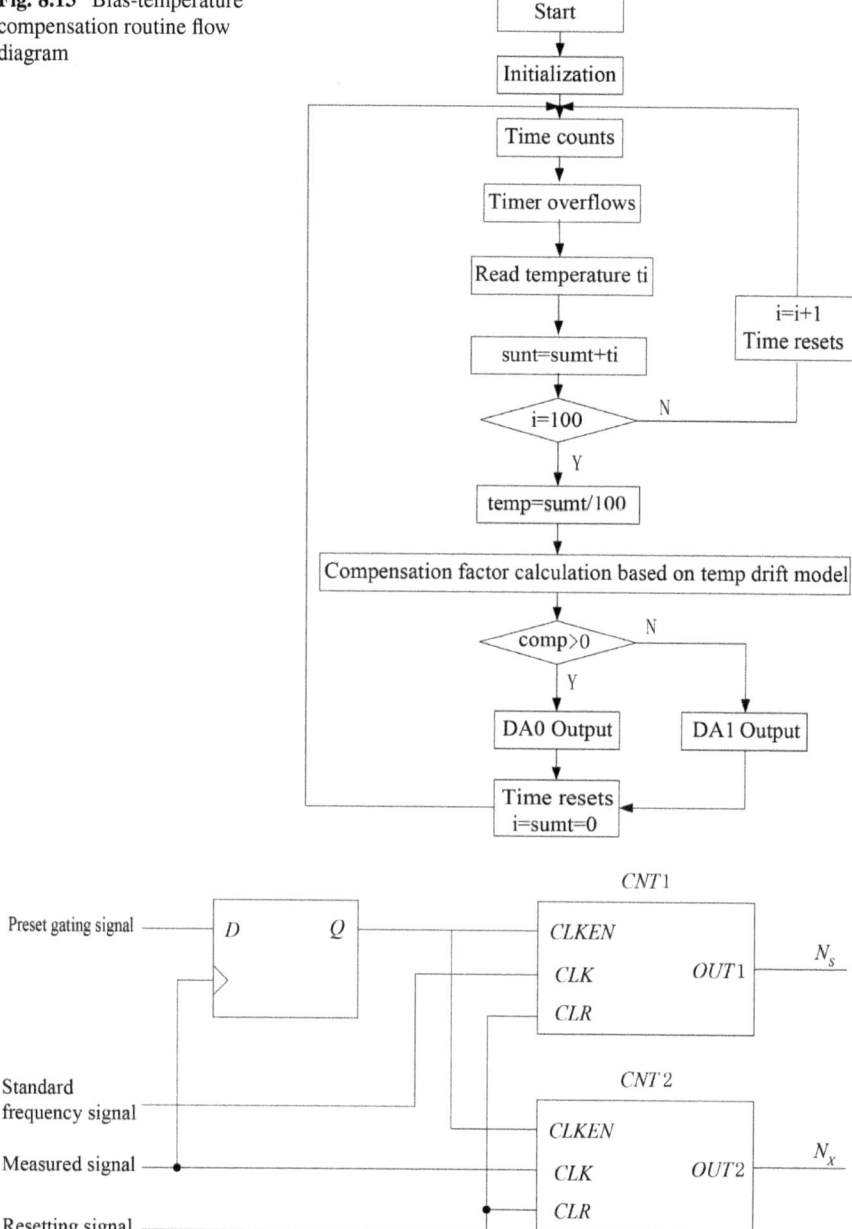

**Fig. 8.14** Functional block diagram for equal-precision frequency measurement

**Fig. 8.15** Signal graph for
equal-precision frequency
measurement method

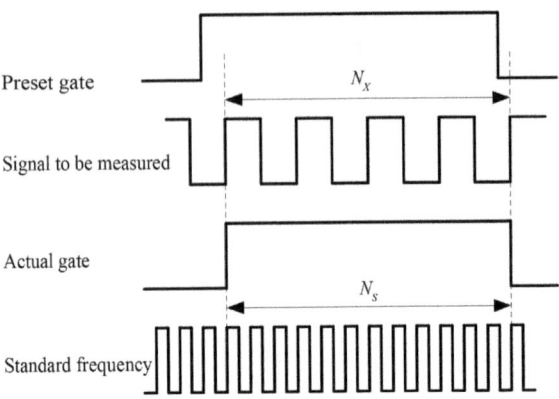

$f_s$ and is input from CLK, the clock input terminal of CNT1. Let the actual frequency
of the measured signal be $f_x$. After rectification and amplification, it is input from
the CLK of CNT2 and simultaneously input into the D trigger. If the preset gating
signal is high, when the measured signal passes through the Q terminal of the D
trigger after rectification, counters CNT1 and CNT2 will be activated. Then, the
two counters simultaneously start counting the standard frequency signal and the
measured signal separately. If the preset gating signal is low, the subsequent rise of
the measured signal will turn off the two counters simultaneously. The actual gate
time is not a fixed value but many times as high as the full period of the frequencies
to be measured, i.e., it is synchronous with the frequencies being measured, as shown
in Fig. 8.15.

In the equal-precision frequency measurement method, the actual gate time is
$N_x/f_x$. If the standard frequency is $f_x$, the gate time can be approximately expressed
as $N_s/f_s$, so,

$$\frac{N_x}{f_x} = \frac{N_s}{f_s} \Rightarrow f_x = \frac{N_x}{N_s} \times f_s \tag{8.18}$$

The measurement accuracy of the equal-precision frequency measurement method
is related to the preset gate width and standard frequency, but is not affected by the
frequency of the measured signal. Therefore, the $\pm 1$ error arising during the counting
of the frequency signals can be eliminated, and an equal-precision measurement can
be performed across the entire frequency band. Extending the preset gate time $T_{pr}$
or increasing the standard frequency $f_s$ can increase $N_s$, reduce measurement error
and enhance measurement accuracy.

If the standard frequency is 10 MHz, the frequency of the signals to be measured
is about 4000 Hz and the preset gate time is set to 1 s, then this method will have a
measurement error as follows

$$\Delta f = \left( \frac{N_x}{N_s} - \frac{N_x}{N_s + 1} \right) f_s = \left( \frac{4000}{10^7} - \frac{4000}{10^7 + 1} \right) \times 10^7 = 4 \times 10^{-4} \, (\text{Hz})$$

$$(8.19)$$

As can be seen, this method gives a very accurate frequency reading, up to $10^{-4}$ Hz under ideal conditions.

(2)    System implementation

### I.    Overall Structure

The measurement circuit is simplified in this system, where the MCU is used alone for the implementation of all functions, including the counting of standard frequency signals, the generation of the preset gating signal, the calculation of the signals, and the calculation of the compensation factor and DA output. C8051F411 MCU forms the core of the entire system. The MCU has a built-in PCA (programmable counter array), which can be used to count the signals to be measured and the standard frequency signals. The accuracy of the standard frequency signal has a direct influence on the result of the frequency measurements. A high-precision temperature-compensated crystal oscillator (TCCO) is used as the standard frequency signal and also as the system clock of the MCU. The basic structure of the system is shown in Fig. 8.16.

### II.    Working Principles

The PCA counter in C8051F411 MCU is used for an equal-precision measurement of frequencies. The PCA counter in C8051F411 MCU consists of a dedicated 16-bit counter/timer and 6 capture/compare modules. Its structure is shown in Fig. 8.17.

There are many options for the PCA counter. For example, a special register value can be set. In this design, the system clock of the MCU is set as an external clock, i.e., a TCCO. The 16-bit PCA counter is in an active state all the time, and the snapshot register can automatically latch the value of the counter register. When reading the value of the 16-bit counter, the operation of the counter is not affected.

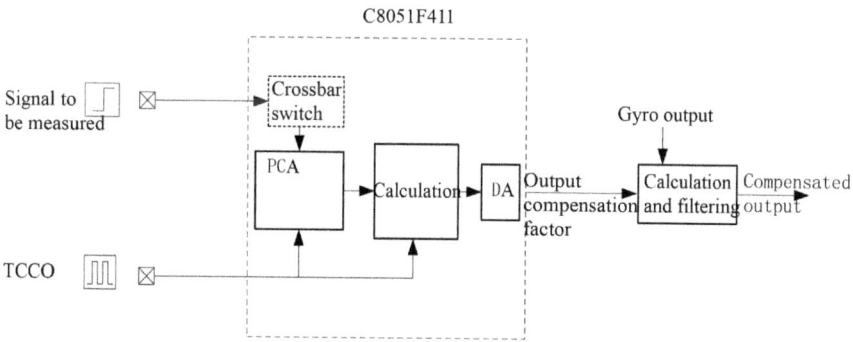

**Fig. 8.16** Bias-frequency compensation system diagram

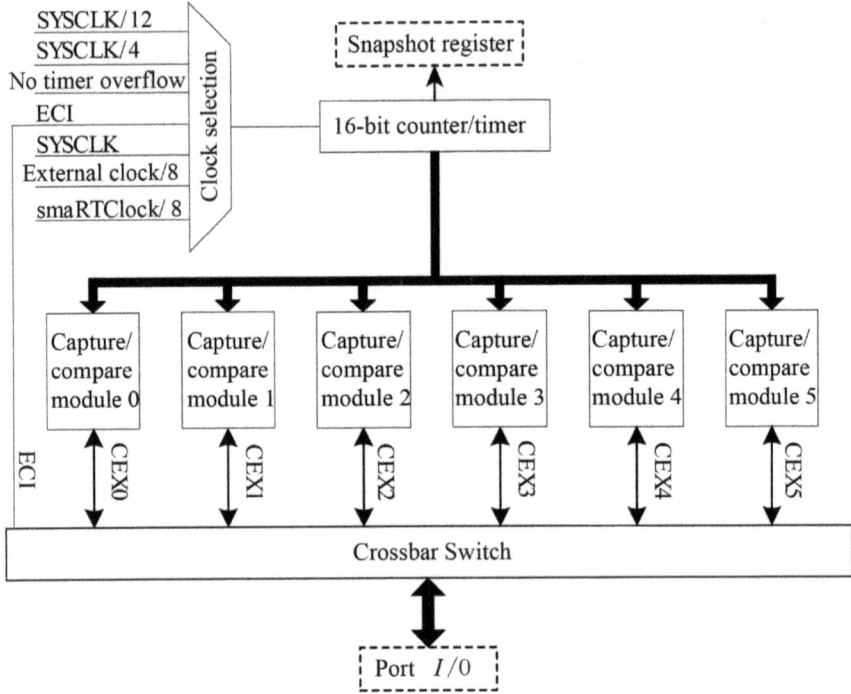

**Fig. 8.17** PCA chart

The capture/compare module of the PCA can work in multiple modes: edge-triggered capture, software timer, high-speed output, frequency output, 8-bit PWM and 16-bit PWM. The capture/compare module works in edge-triggered capture mode. In this mode, the PCA counter value is captured and loaded into the capture/compare registers (PCA0CPLn and PCA0CPHn) of the corresponding modules when any electrical level jumping appears on CEXn. A special register value can be set to determine the rising or falling edge triggers. The principle of the PCA capture is shown in Fig. 8.18.

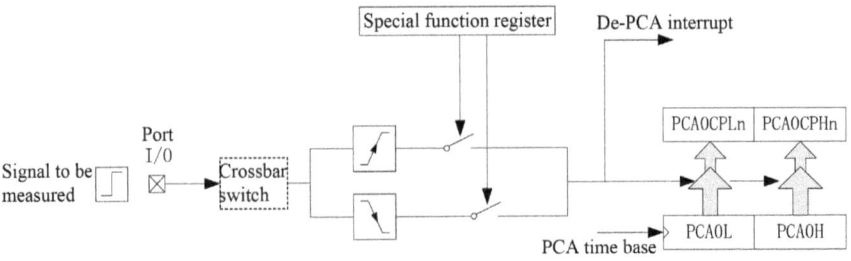

**Fig. 8.18** Schematic diagram of PCA capture

The following is the process in which PCA is used for signal frequency measurement: The PCA capture/compare module is set to edge-triggered capture mode, and the signal to be measured is input through the I/O. When the rising edge of the signal arrives, a PCA interrupt is generated. At the same time, the value of the PCA0 counter register is loaded into the PCA0CP register and read from the PCA interrupt service routine. The PCA time base is selected as the signal of the system clock, namely the TCCO. When the next rising edge of the signal to be measured arrives, another PCA interrupt is generated, and the value of the PCA0 counter is loaded into the PCA0CP register. We can obtain the number of PCA time-base standard frequency signals in one period of the signal to be measured by subtracting the value of the PCA0CP register at the arrival of the last rising edge from the current value of the PCA0CP register. The frequency of the signal to be measured can be calculated by adding together the various D-values of the PCA0CP register, and obtaining the number $N_s$ of standard frequency signals in $N_x$ periods of the signal to be measured.

Let the resonant frequency of the CVG be 4000 Hz. A TCCO of 10 MHz is selected as the standard frequency signal, i.e., the system clock of the MCU. The 16-bit PCA counter has a counting range of 0–65,535, so the PCA counter overflows at most once during one period of the signal to be measured. Since the value of the PCA register is an unsigned number, the direct subtraction method can be used to calculate the number of standard frequencies in one period of the signal.

Let the preset gating time be 1 s. During this period, there are approximately 4000 drive signals at around $10^7$ standard frequencies. An unsigned long integer variable can be defined in the routine to store the number of standard frequencies. Because there is a 32-bit unsigned long integer in the C8051F411 MCU, the maximum value of this variable is $2^{32}-1 \approx 4.3 \times 10^9$. Therefore, it is feasible to set the reset gating time to 1 s.

Frequency measurement error calculation: for equal-precision frequency measurement, the actual gating time is an integer multiple of the signal period to be measured, and there may be an error of $\pm 1$ in the number of standard frequencies. This error is assumed to be 1. Moreover, considering that standard frequency signals come from the TCCO, the frequency will not be completely unchanged. For example, a TCCO of 10 MHz can be selected. Its frequency accuracy in the entire temperature zone is below 2 ppm and maximum frequency variation is 20 Hz, meaning the frequency measurement error caused is as follows:

$$\Delta f = \left( \frac{N_x}{N_s} - \frac{N_x}{N_s + 21} \right) f_s = \left( \frac{4000}{10^7} - \frac{4000}{10^7 + 21} \right) \times 10^7 = 8.4 \times 10^{-3} \, (\text{Hz})$$

(8.20)

### III. Routine Design

After the MCU is powered up, the register and other variables are initialized before the rising edge of a signal to be measured triggers the PCA interrupt. After this, the frequency and compensation factor can be calculated in the PCA interrupt service routine. Figure 8.19 shows the PCA interrupt service routine.

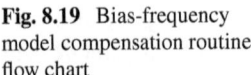

**Fig. 8.19** Bias-frequency model compensation routine flow chart

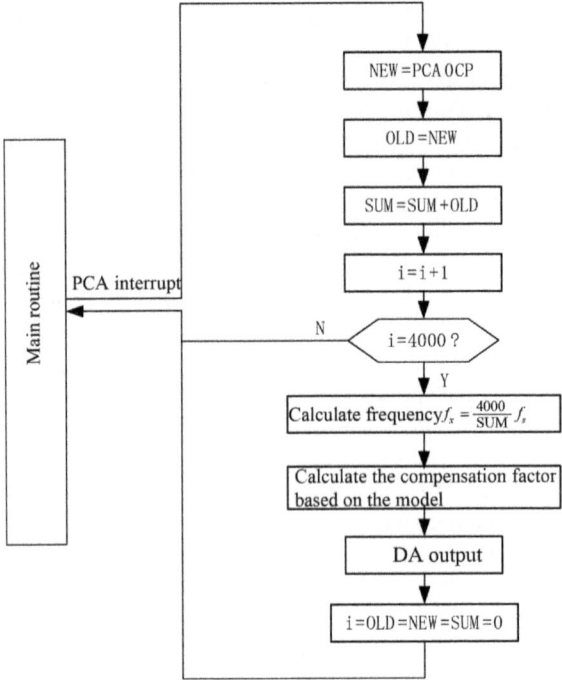

# References

1. Wang, X., Wu, W. Q., Fang, Z., Luo, B., Li, Y., & Jiang, Q. G. (2012). Temperature drift compensation for hemispherical resonator gyro based on natural frequency. *Sensors, 12*(5), 6434–6446.
2. Xiang, X. (2014). *Research on the drift mechanism and suppression technology of the cup-shaped wave gyroscope*. National University of Defense Technology Press.
3. Choi, S. Y., & Kim, J. H. (2011). Natural frequency split estimation for inextensional vibration of imperfect hemispherical shell. *Journal of Sound and Vibration, 330*(9), 2094–2106.
4. Chikovani, V., Yatsenko, Y. A., Barabashov, A., Kovalenko, V., Scherban, V., & Marusyk, P. (2007). Thermophysical parameters optimization of metallic resonator CVG and temperature test results. In *Proceedings of the Petersburg Conference on Integrated Navigation Systems*, St. Petersburg.
5. Yongmeng, Z. (2012). *Research on the temperature characteristics and compensation method of cup-shaped gyroscopes*. National University of Defense Technology.